Diagnostic
Enzymology

Analytical Chemistry by Open Learning

Titles in Series:

Samples and Standards
Sample Pretreatment
Classical Methods
Measurement, Statistics and Computation
Using Literature
Instrumentation
Chromatographic Separations
Gas Chromatography
High Performance Liquid Chromatography
Electrophoresis
Thin Layer Chromatography
Visible and Ultraviolet Spectroscopy
Fluorescence and Phosphorescence Spectroscopy
Infra Red Spectroscopy
Atomic Absorption and Emission Spectroscopy
Nuclear Magnetic Resonance Spectroscopy
X-Ray Methods
Mass Spectrometry
Scanning Electron Microscopy and Microanalysis
Principles of Electroanalytical Methods
Potentiometry and Ion Selective Electrodes
Polarography and Other Voltammetric Methods
Radiochemical Methods
Clinical Specimens
Diagnostic Enzymology
Quantitative Bioassay
Assessment and Control of Biochemical Methods
Thermal Methods
Microprocessor Applications

Diagnostic Enzymology

Analytical Chemistry by Open Learning

Author:
DAVID HAWCROFT
Leicester Polytechnic

Editor:
ARTHUR M. JAMES

on behalf of ACOL

Published on behalf of ACOL, London
by
JOHN WILEY & SONS
Chichester · New York · Brisbane · Toronto · Singapore

Library of Congress Cataloging in Publication Data:

Hawcroft, David M.
  Diagnostic enzymology.

  (Analytical chemistry by open learning)
  Includes bibliographies and index.
  1. Clinical enzymology. I. James, A. M. (Arthur M.),
1923– . II. ACOL (Firm : London, England)
III. Title. IV. Series. [DNLM: 1. Enzyme Tests—
methods—programmed instruction. 2. Enzymes—
diagnostic use—programmed instruction. QY 18 H389d]
RB48.H38 1987   616.07'56   86-28305
ISBN 0 471 91398 7
ISBN 0 471 91399 5 (pbk.)

British Library Cataloguing in Publication Data:

Hawcroft, David
  Diagnostic enzymology.—(Analytic chemistry
  1. Diagnosis, Laboratory   2. Clinical
  enzymology.
  I. Title   II. James, Arthur M.
  III. Analytical Chemistry by Open Learning.
  (Project)   IV. Series
  616.07'56   RB48

  ISBN 0 471 91398 7
  ISBN 0 471 91399 5 Pbk

Printed and bound in Great Britain

Analytical Chemistry

This series of texts is a result of an initiative by the Committee of Heads of Polytechnic Chemistry Departments in the United Kingdom. A project team based at Thames Polytechnic using funds available from the Manpower Services Commission 'Open Tech' Project have organised and managed the development of the material suitable for use by 'Distance Learners'. The contents of the various units have been identified, planned and written almost exclusively by groups of polytechnic staff, who are both expert in the subject area and are currently teaching in analytical chemistry.

The texts are for those interested in the basics of analytical chemistry and instrumental techniques who wish to study in a more flexible way than traditional institute attendance or to augment such attendance. A series of these units may be used by those undertaking courses leading to BTEC (levels IV and V), Royal Society of Chemistry (Certificates of Applied Chemistry) or other qualifications. The level is thus that of Senior Technician.

It is emphasised however that whilst the theoretical aspects of analytical chemistry can be studied in this way there is no substitute for the laboratory to learn the associated practical skills. In the U.K. there are nominated Polytechnics, Colleges and other Institutions who offer tutorial and practical support to achieve the practical objectives identified within each text. It is expected that many institutions worldwide will also provide such support.

The project will continue at Thames Polytechnic to support these 'Open Learning Texts', to continually refresh and update the material and to extend its coverage.

Further information about nominated support centres, the material or open learning techniques may be obtained from the project office at Thames Polytechnic, ACOL, Wellington St., Woolwich, London, SE18 6PF.

How to Use an Open Learning Text

Open learning texts are designed as a convenient and flexible way of studying for people who, for a variety of reasons cannot use conventional education courses. You will learn from this text the principles of one subject in Analytical Chemistry, but only by putting this knowledge into practice, under professional supervision, will you gain a full understanding of the analytical techniques described.

To achieve the full benefit from an open learning text you need to plan your place and time of study.

- Find the most suitable place to study where you can work without disturbance.

- If you have a tutor supervising your study discuss with him, or her, the date by which you should have completed this text.

- Some people study perfectly well in irregular bursts, however most students find that setting aside a certain number of hours each day is the most satisfactory method. It is for you to decide which pattern of study suits you best.

- If you decide to study for several hours at once, take short breaks of five or ten minutes every half hour or so. You will find that this method maintains a higher overall level of concentration.

Before you begin a detailed reading of the text, familiarise yourself with the general layout of the material. Have a look at the course contents list at the front of the book and flip through the pages to get a general impression of the way the subject is dealt with. You will find that there is space on the pages to make comments alongside the text as you study—your own notes for highlighting points that you feel are particularly important. Indicate in the margin the points you would like to discuss further with a tutor or fellow student. When you come to revise, these personal study notes will be very useful.

∏   When you find a paragraph in the text marked with a symbol such as is shown here, this is where you get involved. At this point you are directed to do things: draw graphs, answer questions, perform calculations, etc. Do make an attempt at these activities. If necessary cover the succeeding response with a piece of paper until you are ready to read on. This is an opportunity for you to learn by participating in the subject and although the text continues by discussing your response, there is no better way to learn than by working things out for yourself.

We have introduced self assessment questions (SAQ) at appropriate places in the text. These SAQs provide for you a way of finding out if you understand what you have just been studying. There is space on the page for your answer and for any comments you want to add after reading the author's response. You will find the author's response to each SAQ at the end of the text. Compare what you have written with the response provided and read the discussion and advice.

At intervals in the text you will find a Summary and List of Objectives. The Summary will emphasise the important points covered by the material you have just read and the Objectives will give you a checklist of tasks you should then be able to achieve.

You can revise the Unit, perhaps for a formal examination, by re-reading the Summary and the Objectives, and by working through some of the SAQs. This should quickly alert you to areas of the text that need further study.

At the end of the book you will find for reference lists of commonly used scientific symbols and values, units of measurement and also a periodic table.

Contents

Study Guide

This Unit is concerned with diagnostic enzymology. This important area of analysis in medical laboratories is possible because of the fact that while many cellular enzymes are normally present at low levels in blood and other extracellular fluids, diseases, infections and other illnesses frequently result in an increase in the level. This means that by careful analysis of the activity of appropriate enzymes, the presence, extent and location of a disease can often be determined.

The measurement of the enzyme activity and the interpretation of the results obtained, need to be done carefully and so this Unit begins with a reasonably detailed discussion of the nature of enzymes, the methods used for their measurement and the problems associated with obtaining reproducible and meaningful results.

Since so much introductory material is provided only limited background knowledge is required in order to follow this Unit. To understand the basic concepts a knowledge of chemistry to GCE 'O' level or ONC standard would be useful and to place the Unit in its medical context, a study of animal and micro-biology to GCE 'A' level might be of value. However an intelligent and interested reader should be able to master the whole Unit without undue problems even without this prior study.

The bibliography contains a range of sources of additional information should it be needed, and gives alternative views on subjects which are contentious and uncertain.

The concurrent study of other ACOL Units is not essential, although some of them should be of general interest. In particular the Units in the clinical section that deal with the collection and preservation of clinical specimens, and method evaluation, are of direct relevance. Other Units that deal with separation techniques and reaction measurement will be of value if you wish to extend the coverage of these subjects contained in this Unit.

General Aims of the Unit

The early parts of this Unit will show you some of the general chemical and biological properties of enzymes which are of relevance to their measurement. They will also cover key aspects of the general approaches taken to the measurement of enzyme activities (including the units in which enzyme reaction rates are expressed), and will indicate the problems and merits of various techniques. Some details of important experimental methods for the measurement of the product of enzyme reactions will be described. Later parts should indicate to you some problems regarding the reliability of enzyme assays and the validation of the results obtained. The Unit culminates with a discussion intended to illustrate the considerable importance of enzyme assays in the diagnosis of human illness and the monitoring of treatment and recovery.

References for further information on various topics are given occasionally in the text and full details of each may be found in the list of Specific References at the end of the Unit. Whilst a knowledge of the contents of these references is well beyond the scope of Enzymology covered in this text, they will provide a fascinating insight for students wishing to study the subject further.

Supporting Practical Work

A short course of laboratory work covering the properties and measurement of enzymes, would be of value to a student studying this Unit.

The aims of such a course would be to:

— provide experience in handling biological material and using simple measuring instruments;

— illustrate techniques used to determine enzyme distribution and activity.

The following experiments are suggested as appropriate to the topics covered in this Unit.

(*a*) Fundamental Properties of Enzymes

— The effect of substrate concentration, pH and temperature on amylase activity measured by the dinitrosalicylate reaction of the maltose product.

— The substrate stereospecificity, and inhibition by heavy metal ions, of α-glucosidase using p-nitrophenylglucosides as substrates.

— The kinetics of the acid phosphatase reaction, measured by the Folin reaction for phenol liberated from the p-nitrophenylphosphate substrate.

(*b*) Sample Problems

— The effect of anticoagulants on urease activity

— A comparison of the stability of amylase and creatine kinase (as measured by the change in their reaction rates) following storage at 4 °C and room temperature over several days.

(*c*) Reaction Measurement

— A comparison of the measurement of lactate dehydrogenase by an end-point and a continuous method.

— A comparison of the measurement of amylase by following the appearance of reducing sugar products and the disappearance of the starch substrate.

(*d*) Diagnostic Enzymology

— The separation of LDH isoenzymes by cellulose acetate or agarose electrophoresis and their visualisation by an *in situ* formazan reaction.

— The asymmetry of distribution of aminotransferases among the major body organs, measured using a commercial kit assay based upon the dinitrophenylhydrazine reaction for the keto-acid products.

Bibliography

A wide range of textbooks is available and they vary in the following main ways:

— their depth of coverage (roughly indicated by their thickness and their price);

— the extent to which they extend from central aspects of biochemistry into:

  – cellular aspects,
  – molecular biology ie genetics,
  – aspects of physical chemistry,
  – experimental techniques,
  – medical applications.

The following is a representative selection from the many books available – you are not expected to read them all(!) and may well find others that are just as suitable as these.

1. BIOCHEMISTRY

Ottaway J. and Apps D. *Biochemistry*, 4th edition, Ballière Tindall 1984. Very good coverage at intermediate level.

Gareth-Norris J. *A Biologist's Physical Chemistry*, 2nd edition, Arnold 1974. Gives a good coverage of reaction kinetics, energy changes, etc.

Stryer L. *Biochemistry*. Freeman 2nd edition 1981. Readable, well illustrated and comprehensive.

Lehninger A. *Principles of Biochemistry*. Worth 1980. Comprehensive and authoritative, but less detailed than his related book *Biochemistry*.

2.   GENERAL CELL BIOLOGY

Hopkins C. *Structure and Function of Cells*, Saunders 1978. An introductory discussion of the structure of cells and the function of the cell components.

Albertis B., Bray D., Lewis J., Raff M., Roberts K., and Watson D., *Molecular Biology of the Cell*, Garland 1983. A detailed discussion of cells which stretches into aspects of biochemistry and genetics.

Avers C. J. *Molecular Cell Biology*, Addison-Wesley 1985. Similar to Albertis.

3.   GENERAL ASPECTS OF ENZYMOLOGY

Price N. C. and Stevens L. *Fundamentals of Enzymology*, Oxford University Press 1982. An introductory text.

Fersht A. *Enzyme Structure and Mechanism*, Freeman 1984. Includes some useful material on genetic engineering.

Bergmeyer H. U. (ed) *Principles of Enzymatic Analysis*, Verlag Chemie 1978.

Methods in Enzymology, more than 120 volumes, Academic Press. Very comprehensive and authoritative coverage of analytical methods for specific enzymes, specialised techniques etc.

Acknowledgements

Fig. 7.2a from Tietz N. (1983) *Clinical Chemistry*, **26**, 751–61. Reproduced with permission from American Association for Clinical Chemistry.

Data in Figs. 2.1a and 7.2b from Wellcome Group Quality Control Programme. Permission requested.

Fig. 9.3c,d,e reproduced from *Cardiac Infarction?* with permission from E. Merck GmbH.

1. The Enzymes: Introductory Background

Overview

This part of the Diagnostic Enzymology Unit will consider some of the fundamental features of enzyme catalysis of relevance to their measurement in biological samples.

1.1. METABOLISM

It is possible to debate endlessly about philosophical subjects such as 'What is an organism?' and the topic is of course of interest to educated non-scientists, including theologians and humanists, as well as to theoretical biologists. While many of these debates centre around the existence and nature of 'intelligence', 'the soul' etc, at a much more fundamental level organisms can be considered merely as a complex series of inter-locked chemical reactions. The many thousands of these reactions that go to make a living organism are arranged into branched chains called pathways and are collectively called the metabolism. The object of metabolism is to obtain energy and basic building materials, which in our case come from the food we eat, and to convert these materials into the more complex chemicals we require to build and maintain our body structure, and to carry out various processes such as nerve conduction, muscle activity etc. Details of the organisation of metabolism and descriptions of the pathways and their control can be found in most general biochemistry textbooks including those given in the reference list.

1.2. ENERGETICS OF REACTIONS

Now it is the case that for the vast majority of metabolic reactions the substrates do not spontaneously convert into the products at an appreciable rate and it is worth exploring very briefly why this is the case. You can regard any mixture of substrate molecules as having an average energy value at the start of a reaction and similarly the molecules of the product mixture would also have an average energy value at the end. At first sight you might expect that for those reactions where the latter value is less than the former then a spontaneous conversion should occur (Fig. 1.2a), and then all your body proteins would simply separate into their constituent amino acids!!

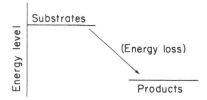

Fig. 1.2a. *Energy levels in a simple reaction*

Obviously this does not happen because it ignores some problems arising from the nature of the chemical changes which occur when the substrates convert into the products. Without going into great detail of the relevant chemistry it can simply be said that in order to generate the product the substrates need to pass through so-called intermediate and transition states where the actual atom movements and bond rearrangements take place.

Now these states are relatively unstable and of high energy and as a consequence an 'energy hump' appears in the profile which means that a certain amount of energy known as the activation energy (E_a) has to be acquired by the substrate molecules before they can form products (Fig. 1.2b). For those reactions which are to a certain degree spontaneous sufficient energy can be obtained from the environment to surmount their relatively low activation energy barrier.

α- and β-glucose provide interesting biological examples of this process – each can be synthesised and isolated but solutions cannot be
stored or purchased separately since they spontaneously interconvert to a mixture of the two forms.

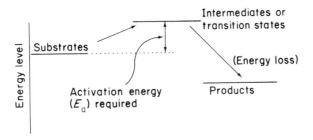

Fig. 1.2b.   *Energy levels of the components of a simple reaction.*

∏   Examine the following profiles of the energy changes occurring during some hypothetical reactions.

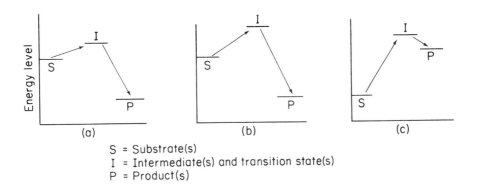

S = Substrate(s)
I = Intermediate(s) and transition state(s)
P = Product(s)

Which reaction do you think will proceed most readily and
why?

Naturally a lot of factors can influence reaction rates but if
we restrict ourselves to considering just the relative energy
levels of substrates and products and the activation energy
requirements we can say the following.

(*a*) This is likely to proceed most readily since on the one hand the products have less energy than the substrates ie, the reaction is 'downhill', and in addition the activation energy is quite low.

(*b*) While this is also downhill, the activation energy requirement is much greater and the reaction will therefore proceed less readily.

(*c*) This reaction is 'uphill' ie, the energy of the products is greater than that of the reactants, and an actual energy input is required. It might be possible to gain this from the environment. In a biological situation the reaction would have to be directly coupled to one yielding sufficient free energy to 'drive' the reaction.

If you take a sample of molecules not all of them will possess exactly the same energy at any particular time, molecules gain and lose energy as a result of collisions. The distribution of energy among molecules is represented by the solid line on the graph in Fig. 1.2c and if the molecules require an activation energy represented by line X in order to react, then none will be able to react. However, if they require an energy of only Y, then some will be able to react.

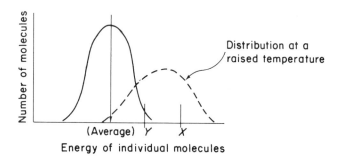

Fig. 1.2c. *Distribution of energy among molecules*

∏ Can you think of two distinct ways of changing the energetics of the situation so that more molecules would react? Consider this as a simple theoretical problem for now and do not worry about the practicalities of achieving what you decide.

Now it is possible to increase the reaction rate by raising the environmental temperature in order to increase the number of molecules with energy of more than X; which would give a distribution shown by the dashed line in Fig. 1.2c. It is commonplace in chemical laboratories, industry and even domestically to do this by heating; but this is not a method readily available to living organisms. A completely different approach is to reduce the activation energy by the use of a catalyst, and while this is an approach also used industrially and in laboratories, it is the method that organisms have developed to a high degree of sophistication. The catalysts in question are of course the enzymes which are the subject of this Unit.

According to the Arrhenius equation, the rate constant, k, of a reaction is given by:

$$k = A \exp\left(-E_a/RT\right)$$

where A is a constant depending on the collison number and the term $\exp\left(-E_a/RT\right)$ is a measure of the number of molecules with energy E_a.

Thus if $E_a = 115$ kJ mol^{-1} at 310 K, then the term $(-E_a/RT)$ is equal to 4.2×10^{-20}. If E_a is reduced to half this value (eg as a result of the presence of a catalyst) the value of the exponential term rises to 2×10^{-10} giving a marked increase in the rate of reaction.

Enzymes achieve this dramatic effect in a number of ways including changing the chemistry of the reaction to produce a lower energy route to the product, which is much like finding a pass through a mountain range. In other words by using different intermediates and transition states they reduce the activation energy requirement for the reaction (Fig. 1.2d).

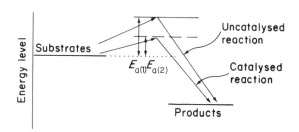

Fig. 1.2d.   *The effect of a catalyst on the energy change occurring during a reaction. $E_{a(2)}$ (catalysed reaction) $<$ $E_{a(1)}$ (uncatalysed reaction)*

∏   The activation energy for the conversion of hydrogen peroxide to water and oxygen at constant temperature is (*a*) 75
240 J mol⁻¹, for the uncatalysed reaction, (*b*) 50 160 J mol⁻¹
in the presence of platinum powder, and (*c*) 8 360 J mol⁻¹
in the presence of catalase. Which reaction do you think will
proceed the fastest?

Assuming that activation energy is the sole or at least the major determining factor then the reaction will proceed fastest
when catalase is present. The extent to which catalase can
lower the activation energy barrier (about 89%) shows dramatically how effective enzymes are as catalysts and to quote
Stryer (1981) many enzymes have 'attained kinetic perfec-
tion' by which he means that they can work so fast that their
rate of reaction is limited solely by the diffusion rate of the
molecules involved. The value for the uncatalysed reaction is
sufficiently high for hydrogen peroxide to be stable at room
temperature because environmental heat cannot supply sufficient energy for any of the molecules to overcome the activation energy barrier.

SAQ 1.2a   Which of the following statements is correct? The activation energy of an enzyme reaction is:

(*i*)   the energy needed to activate the enzyme for reaction;

(*ii*)   the energy needed to activate the substrate for reaction and form the transition or intermediate states;

(*iii*)   the energy liberated during the reaction;

(*iv*)   the energy retained by the products;

(*v*)   the energy provided by the products.

SAQ 1.2b   The following is a diagram representing the energy changes occurring during the reaction.

$$H_2 + 0.5\,O_2 \rightarrow H_2O$$

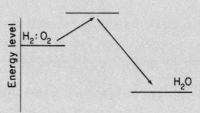

Draw in the curve representing the changes occurring in the presence of platinum as a catalyst.

SAQ 1.2c

The following table shows the activation energy for the hydrolysis of fructose-containing polymers:

sucrose + H_2O → glucose + fructose

raffinose + $2H_2O$ → glucose + 2 fructose

Reaction (hydrolysis of)	Catalyst	Activation energy kJ mol^{-1}
(*i*) sucrose	H^+	108.7
(*ii*) sucrose	malt sucrase	54.4
(*iii*) sucrose	yeast sucrase	46.0
(*iv*) raffinose	yeast sucrase	46.0

1. Which of these reactions (*i*) and (*ii*) is likely to proceed more readily?

2. Note the difference in energy value between (*ii*) and (*iii*) and the similarity of value for (*iii*) and (*iv*). Are these data suggestive of the enzyme, or the substrate and the particular reaction being the major determinant of the activation energy value?

1.3. ENZYMES AS CATALYSTS

This Unit is concerned with the analysis of enzymes and some areas where the results of this analysis are of interest, and it is not really relevant for us to consider the mechanisms by which enzymes carry out their catalytic role. However, some properties and features of enzymes are important because of the way they affect analytical methods or the interpretation of the results.

1.3.1. Numbers of Enzyme Molecules

Perhaps one of the most obvious and fundamental factors is that they are catalysts and as such are recycled. This, together with their great efficiency (see later) allows cells to manage with comparatively small numbers of enzyme molecules and hence save the space, materials and energy which would be necessary if this were not the case. The small numbers of molecules of each enzyme present in biological specimens could present us with problems regarding their assay but again the points just mentioned work in our favour, because these highly efficient catalysts are able to generate considerable quantities of product molecules for us to measure. In any case if necessary we can simply allow the reaction to proceed for longer so that the recycling catalyst can continue to generate product until sufficient is available for analysis. These properties are also of value in the use of enzymes as reagents and this will be dealt with in another Unit.

1.3.2. The Effectiveness of Enzymes as Catalysts

A second characteristic of great significance is one we have alluded to already, that is their incredible effectiveness as catalysts. Inorganic catalysts such as platinum are fairly simple structurally and as such have simple mechanisms of action, but cells can produce the more complex enzyme catalysts in an almost infinite variety of compositions and structures. This in effect allows them to tailor-make an enzyme for a particular reaction. They can produce molecules of staggering catalytic effectiveness and it is common for enzymes to increase reaction rate by 10^8 to 10^{15} fold and be capable of converting thousands of substrate molecules per second. It is important to note in passing that as with non-enzymic catalysts, enzymes have no effect on the equilibrium position of the reaction, merely on the time taken to reach it.

1.3.3. Enzyme Inhibition

Unfortunately the sheer complexity of enzyme structure and mechanism means that enzymes can very easily be interfered with and serious problems with the reliable determination of enzyme reaction rates can arise. The observed reaction rate can change due to the influence of factors such as substrate concentration, time of incubation of the reaction mixture, environmental temperature and pH. In addition a relatively wide spectrum of substances can act as inhibitors of enzyme activity generally, but not always, by binding to the site where the enzyme carries out its reaction (the so-called active site). These include metal ions (which may be present in water supplies), certain anions (including the common anticoagulant fluoride) and competitive inhibitors which are compounds with structures similar enough to the normal substrate to compete with it, bind to the enzyme active site and then effectively block it by failing to react.

Π   The enzyme succinic dehydrogenase is involved in the Krebs' or Citric Acid Cycle of aerobic respiration and catalyses the oxidation of succinic acid as shown below.

$$
\begin{array}{ccc}
\text{COOH} & & \text{COOH} \\
| & & | \\
\text{CH}_2 & \longrightarrow & \text{CH} \\
| & & \| \qquad + \ 2\text{H}^+ \ + \ 2\text{e}^- \\
\text{CH}_2 & & \text{CH} \\
| & & | \\
\text{COOH} & & \text{COOH}
\end{array}
$$

Succinic acid   Fumaric acid

Which of the following acids do you think could act as competitive inhibitors of this enzyme?

			COOH	COOH
COOH			CHOH	CH$_2$
	COOH	COOH	CHOH	CH$_2$
CH$_2$	CH$_3$	COOH	CHOH	CH$_2$
COOH			CHOH	COOH
			CH$_2$OH	
(a)	(b)	(c)	(d)	(e)
Malonic acid	Ethanoic acid	Oxalic acid	Glucuronic acid	Glutaric acid

In this case it is likely that the substrate attaches to the enzyme surface by the two carboxyl groups and hence (*a*), (*c*) and (*e*) are quite good competitive inhibitors. In each case the type, arrangement or absence of groups between the carboxyl groups makes it impossible for the enzyme to carry out its normal dehydrogenation reaction.

The other compounds (*b*) and (*d*) do not have the requirements for attachment to the enzyme and hence cannot act as inhibitors in this way.

1.3.4. Enzyme Denaturation

Changes in the environment surrounding enzymes can also affect their stability quite dramatically and salts and solvents commonly cause the molecules to associate together and hence precipitate, rather than remain separate and active. A superficially similar, but in fact quite different, process is denaturation in which a normally permanent change in three dimensional structure causes not only a loss of activity (hence 'de-nature') but often a loss of solubility also. Most of the situations in which denaturation occurs are a consequence of the addition of energy to the molecule which is why enzymes are sensitive to temperature, vigorous shaking, low wavelength radiations and even strong sunlight. Chemical changes resulting from pH extremes in unbuffered environments, or detergents (which commonly cling to laboratory glassware) also frequently cause irreversible denaturation. Many of these situations can arise during clinical investigations and the most significant consequence of both precipitation and denaturation is a loss of catalytic activity which is all the more serious if it is partial and goes unnoticed.

SAQ 1.3a

The enzyme urease catalyses the hydrolysis of urea to carbon dioxide and ammonia.

$$(H_2N)_2CO + H_2O \rightarrow 2NH_3 + CO_2$$

The compounds thiourea $(H_2N)_2CS$, hydrazine $(H_2N)_2$ and semicarbazide $H_2N.NH.CO.NH_2$ all give an inhibition in excess of 95% whereas acetamide $H_2N.CO.CH_3$ barely inhibits at all.

(*i*) Examine the structures of these molecules carefully and explain why acetamide is a non-inhibitor.

(*ii*) What type of inhibition are these compounds exhibiting?

(*iii*) Why are studies on the structure of inhibitors useful in investigations of enzyme reaction mechanisms?

SAQ 1.3a

1.3.5. Non-enzymic Components of Enzyme Catalysed Reactions

While we have purposefully avoided going into detail of enzyme reaction mechanisms it is important to note that these mechanisms can be very complex and as a consequence enzymes frequently make use of non-proteinaceous materials in their reactions. These are of various types principally:

— prosthetic groups, eg haem, which are covalently attached to the protein and are often involved in the electron movements in the reactions;

— ions eg Mg^{2+}, Cl^- etc, which have a wide range of roles including acting as catalysts in their own right and as activators of enzyme molecules;

— coenzymes eg nicotinamide adenine dinucleotide (NAD^+), which are free, chemically complex, organic compounds derived from vitamins and which act as shuttles carrying molecular components such as electrons, hydrogen atoms, and amine, methyl and other groups between various coupled reactions.

∏ The function of coenzymes in carrying groups within metabolism is well illustrated by the role of pyridoxal in the metabolism of amino acids. In the equation below $A-NH_2$ represents a hypothetical amino acid.

$$
\begin{array}{ccccccc}
A & & Pyridoxal & & A & & Pyridoxamine \\
| & + & | & \longrightarrow & || & + & | \\
(NH_2) & & (CHO) & & (O) & & (CHNH_2)
\end{array}
$$

You can see that the pyridoxal molecule has an exposed aldehyde group which can pick up the amine group removed from the substrate $(A-NH_2)$ by the enzyme. The pyridoxamine product can donate this amine group in a subsequent reaction.

Bearing in mind the relative scarcity of dietary vitamin molecules and hence coenzyme molecules derived from them, what do you think must happen within cell metabolism to ensure a steady supply of coenzymes molecules?

To maintain a balanced metabolism in which all reactions can proceed when required and with a variable rate it is essential that an adequate supply of coenzyme molecules is maintained. To do this it is necessary to have balancing reactions so that coenzyme molecules converted to a particular derivative by a given reaction can be converted back again for re-use. In some cases these couplings are metabolically close together and effectively are part of a large reaction system. One such system occurs in the transamination process and has been illustrated in part above. The full reaction is shown below.

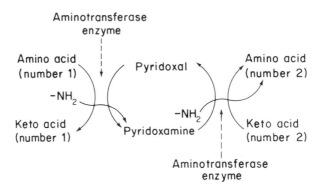

Hopefully you can see from the scheme that in this coupled system an amine group has been exchanged between amino acid 1 and keto-acid 2 in order to create another amino acid (number 2), pyridoxal is used as a carrier and is regenerated.

However occasionally cells have to include reactions of no real value or even ones which are potentially harmful in order to ensure this recycling can occur, and the classic example of this comes from the breakdown of glucose under oxygen free conditions (anaerobic respiration). The reactions are usually called glycolysis and result in the conversion of the coenzyme NAD^+ to its reduced form, NADH. In order to re-generate NAD^+, cells have included fermentation reactions in which potentially lethal compounds such as ethanol and lactic acid are produced but which lead to the concurrent oxidation of NADH to NAD^+ thus allowing the system to operate continuously.

$$NAD^+ + 2H^+ + 2e^- \xrightleftharpoons[\text{Oxidation}]{\text{Reduction}} NADH + H^+$$

SAQ 1.3b

Select from the following list those feature(s) that apply to the prosthetic groups involved in some enzyme reactions.

Prosthetic groups

(*i*) can be distinguished from coenzymes by their solubility in the cell cytoplasm and other aqueous systems;

(*ii*) are normally proteins;

(*iii*) are present at the active site and covalently attached there;

(*iv*) are intimately involved in the enzyme reaction;

(*v*) carry groups from one reaction to another (ie from one enzyme to another);

(*vi*) are inorganic ions.

There are perhaps two main consequences to us of this requirement for additional components, firstly that it increases the number
of ways enzyme reactions can be interfered with or inhibited and
secondly that during measurement one must be careful to include
in the assay mixture an appropriate, non-limiting, concentration of
such required material. It is also necessary to ensure that they are
not lost during enzyme isolation and purification procedures or during a process such as dialysis, which is commonly employed in automatic analysers.

1.3.6. Enzyme Specificity

A final point about enzymes which is of great significance is their
specificity. Enzymes, without proven exception, carry out just a single type of reaction on their substrate and in many cases can also be
said to be absolutely specific as to their substrate. Thus an enzyme
might metabolise the amino acid proline, but not the very similar
hydroxyproline. This commonly extends as far as metabolising only
one isomer of an isomeric compound, due to the inability of the
alternative isomer(s) to fit into the enzyme active site as a result of
having a different shape. Such specificity is of course essential in
the cell in order that metabolism can be controlled. You can imagine the chaos which would result from having molecules capable
of catalysing many different reactions running loose throughout the
cell! This specificity is also of course of great value to us in enzyme
measurement since it allows us to measure the activity of the enzyme
in the presence of many others, simply by supplying its particular
substrate upon which the others generally fail to react. It means that
we do not have to isolate the enzyme of interest before analysis with
all the time, expense and analytical error that would incur.

However not all enzymes have this absolute specificity, some are
employed by the cell as more general purpose enzymes and simply
require the substrate to have certain groups (ie they show 'group
specificity'). A good example of relevance to us in medical laboratory work is the phosphatase type of enzyme which is able to remove
phosphate from a very wide range of compounds as shown below.

$$R-O-PO_3^{2-} + H_2O \rightarrow R-OH + H^+ + PO_4^{3-}$$

Fig. 1.3a illustrates some of the reactions that have formed the basis of assay methods for alkaline phosphatase. Phenylphosphate, nitrophenylphosphate and glycerophosphate are acting as $R-O-PO_3^{2-}$ in these methods.

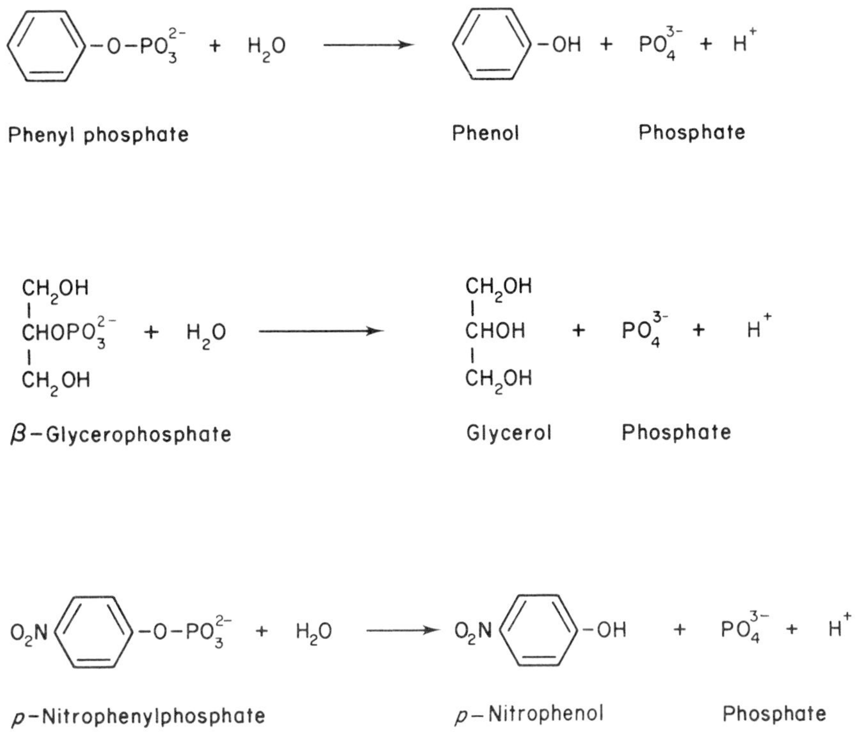

Phenyl phosphate — Phenol — Phosphate

β-Glycerophosphate — Glycerol — Phosphate

p-Nitrophenylphosphate — p-Nitrophenol — Phosphate

Fig. 1.3a. *The use of various substrates in the assay of alkaline phosphatase*

The presence of group specificity in enzymes can raise some analytical problems but has the great benefit that artificial substrates which yield an easily measured product can often be used in their measurement.

∏   What type of specificity is illustrated by each of the following?

(*a*) The activity of alcohol dehydrogenase on a variety of alcohols having low relative molecular mass.

$$XCH_2OH \rightarrow XCHO + 2H^+ + 2e^-$$

(*b*) The activity of urease on urea only, no activity with for example thiourea $CS(NH_2)_2$.

$$CO(NH_2)_2 + H_2O \rightarrow CO_2 + 2NH_3$$
urea

(*c*) the formation of L-glycerol-3-phosphate from glycerol by glycerokinase, with none of the D-isomer produced.

glycerol + ATP → L-glycerol-3-phosphate + ADP.

(*a*) This is a straightforward case of *group specificity* where the enzyme only requires an accessible primary alcohol group ($-CH_2OH$) and this could be attached to a range of other groups. Do you think the enzyme could metabolise the alcohol if it were part of a serine molecule, ($HOCH_2CH(NH_2)COOH$), involved in a protein structure? (See the end of this discussion for the answer to this problem).

(*b*) Urease is showing *absolute specificity*. The properties of the elements oxygen and sulphur are very similar and many oxygen-containing compounds have sulphur analogues, which frequently have some properties in common. However the inability of urease to use the analogue thiourea (probably due to loss of charge on the oxygen of urea) shows a high degree of substrate specificity.

(*c*) Glycerol is not a molecule which is stereoisomeric but the addition of a phosphate group to it creates a chiral centre at carbon atom 3 and so 2 steroisomers of the product, glycerol phosphate, exist. The enzyme active site will have a shape such that the addition will occur only in one particular orientation, thus generating only 1 of the isomers. *Stereospecificity* is therefore the answer here.

$$
\begin{array}{c}
CH_2OH \\
| \\
CHOH \\
| \\
CH_2OH
\end{array}
\quad + \quad PO_4^{3-} \quad \longrightarrow \quad
\begin{array}{c}
CH_2OH \\
| \\
CHOH \\
| \\
CH_2O-P
\end{array}
$$

Now a chiral centre

With regard to the additional question concerning the metabolism of serine and posed in (*a*) above, the answer is that the enzyme probably could not metabolise the alcohol group on the amino acid since it is likely to be inaccessible to the enzyme molecule. This illustrates a point of importance in chemical, but more especially in enzymic, reactions where the catalysts are such large molecules, ie molecular orientation, accessibility and steric hindrance can have profound effects on the rates of reaction.

Occasionally these differences in substrate specificity can be turned to our advantage in medical laboratory work and one important example of this is found in the investigation of the enzyme changes following a myocardial infarction (or 'heart attack'). This topic will be dealt with at length later but it is the case that the enzyme lactate dehydrogenase is present in a number of variants (isoenzymes) of which the so-called type 1 is found in high concentration in the heart. As a consequence of this a dramatic rise in serum concentration of the type 1 is found following a heart attack and it is therefore useful to be able to measure it. There are a variety of methods for the measurement, one of which makes use of the fact that the type 1 isoenzyme, in contrast to the other 4 types, is not absolutely substrate specific. It will metabolise not only lactate but also the methyl derivative (β-hydroxybutyrate), thus allowing us to measure it independently of the others.

SAQ 1.3c

Enzymes such as alkaline phosphatase can re-move phosphate from a wide range of substrates such as those illustrated schematically below. P represents phosphate and the other letters a range of possible chemical structures.

A—P   A—B—P   A
                              \
                               B—P
                              /
                             C

— What is the name given to this type of specificity?

— Why might the enzyme be unable to react with a substrate of the following form?

A   B
  \   /
    P
  /   \
D   C

SAQ 1.3d

If a compound which has stereo-isomeric forms is synthesised chemically then both isomers tend to be produced giving a so-called 'racemic' mixture. How might enzymes be used to isolate just one of these forms?

SAQ 1.3e

What might the benefit be to a pathogenic microorganism in having a relatively high proportion of uncommon stereo-isomers of amino acids in its cell wall? Consider the ways in which the host might try and destroy this pathogenic organism.

SAQ 1.3f

This SAQ is perhaps a rather more difficult problem but think carefully about it because it does highlight the far-reaching consequences of the stereospecificity shown by enzymes.

The following equation shows a simplified version of an important reaction occurring in aerobic respiration, that is the conversion of citric acid to 2-oxoglutaric acid.

$$
\begin{array}{ccc}
\text{COOH} & & \text{COOH} \\
| & & | \\
\text{CH}_2 & & \text{CH}_2 \\
| & & | \\
\longleftarrow\text{-----HO--C--COOH-----}\longrightarrow \qquad\longrightarrow & & \text{CH}_2 \quad + \text{ CO}_2 \; + \; 2\text{H} \\
| & & | \\
\text{CH}_2 & & \text{C} = \text{O} \\
| & & | \\
\text{COOH} & & \text{COOH}
\end{array}
$$

Citric acid   2–Oxoglutaric acid

If you look at the formula of citric acid it would appear that the molecule is symmetrical on either side of the dashed line drawn across it. However subtle radioisotope labelling experiments have shown that the enzyme actually metabolises only one particular end of the molecule. Why do you think this so-called 'Ogston Effect' occurs? Bear in mind that it is thought that the enzyme attaches to three groups in the citric acid molecule and you might like to sketch out the possible orientations of the citric acid molecule on an enzyme surface represented like this. $\longrightarrow$

SAQ 1.3f
(cont.)

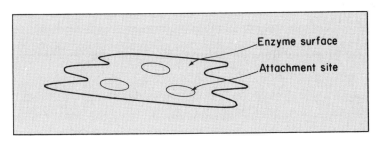

Summary

The fundamental characteristics of enzymes as catalysts have been discussed; in particular their effect on the energetics of reactions, their efficiency, their specificity, enzyme inhibition and denaturation, and the requirement for non-protein components in some enzyme reactions.

Objectives

You should now be able to:

● describe in simple terms the energetics of a typical simple chemical reaction and discuss the effects of a catalyst on this;

● compare the effectiveness of enzymes with other catalytic agents;

● outline the problems generated by the size of enzyme molecules and the requirements for other complex materials in enzyme reactions;

● describe with examples various types of enzyme substrate specificity, discuss the problems specificity can generate, and its usefulness in enzyme assays and clinical diagnosis.

2. Fundamental Aspects of the Measurement of Enzyme Activity

Overview

This part of the Unit will consider some fundamental aspects of the measurement of enzymes especially the concept of the measurement of activities rather than molecular concentrations and the calculation and expression of these activities.

2.1. THE CONCEPT OF THE MEASUREMENT OF ENZYME ACTIVITY

Over the last few decades an enormously diverse range of methods has evolved for measuring molecules of biological interest and a common characteristic among the vast majority of them is that they measure the *quantity* of material in the sample and hence if you know the sample volume you can determine the concentration – which is of course the object of the exercise.

However for some types of molecules this is not the case at all and we in fact measure their *biological activity*, ie their ability to do something of direct relevance to the organism rather than to participate in a general chemical reaction. This is very useful because of course organisms actually depend upon these activities for their survival and the information obtained can say a great deal about the well-being of an organism. Among these compounds are antibodies and enzymes, both of which could be measured by standard protein assays, but rarely are, because their great chemical similarities would make it very difficult to distinguish individual types. Vitamins and hormones are in fact sometimes measured chemically because of their structural diversity, but on many occasions are measured as activities using techniques collectively called bioassays. The value of this bioassay approach is dramatically illustrated by a group of related Israeli families where the isolation, purification and chemical estimation of their growth hormone, or its measurement by immunological methods shows them to have normal concentrations, but something is clearly wrong because they are all dwarfs. Bioassay studies show very poor growth hormone responses, suggesting that they are producing defective molecules with a very low activity.

So, with regard to enzymes we attempt to measure their biological activities and not their concentrations, and as well as being of some philosophical interest and direct biological relevance this is important in that it raises some practical problems. The need to avoid interference with the enzyme activity by inhibitors, or loss of activity due to spontaneous molecular changes, and the development of optimum reaction conditions are among these. Collectively they make enzyme assay an altogether more complex business than simply maintaining the chemical integrity of glucose for example, while carrying out a straightforward chemical reaction for its measurement. One consequence of these problems is that the analytical reliability of enzyme assays is not as good as for more simple materials such as electrolytes or small organic molecules. Whereas the intra-laboratory coefficient of variation for the assay of aminotransferases (transaminases) is often about 7%, the inter-laboratory value can be as high as 20% even in a country such as Germany, where most of the laboratories use the same standardised technique. The data in Fig. 2.1a are taken from a quality control survey and illustrate quite well the wide range of result values obtained for enzyme assays.

The data show the coefficients of variation of the December 1980 results for the analysis of the standard serum supplied to participants in the quality control programme. The number of laboratories carrying out the analyses varied with analyte and ranged from 246–890. For most analytes a range of methods was used. The abbreviations used in (B) are explained in Table 5.

	Analyte		National coefficient of variation (%)
(A)	Albumin		7.3
	Bicarbonate		9.3
	Bilirubin		16.0
	Calcium		3.4
	Chloride		2.4
	Cholesterol		9.3
	Cortisol		25.6
	Creatinine		14.1
	Digoxin		21.3
	Glucose		7.8
	Iron		12.4
	Total Iron Binding Capacity		14.2
	Lithium		13.3
	Magnesium		9.1
	Phosphate		7.3
	Potassium		2.6
	Total Protein		3.8
	Sodium		1.3
	Thyroxin		19.9
	Triacylglycerols		20.8
	Urea		7.5
	Uric Acid		8.8
		Mean	10.8
(B)	ALT		26.8
	AP		42.7
	AM		34.9
	AST		23.8
	CK		32.3
	GGT		24.4
	HBDH		28.8
	LDH		43.8
		Mean	32.2

Fig. 2.1a. *Data from a Wellcome Group Quality Control Programme Report December 1980*

Π   As large molecules, enzymes can induce the production of
      antibodies and as an alternative to chemical assay one could
      determine them by various methods relying on antigen (in
      this case the enzyme) and antibody interaction. Do you think
      the results obtained from this approach would be of value in
      determining enzyme activities?

      The answer is no, while the reaction with antibody is one with
      an important biological basis and implications, it has no re-
      lationship with the enzyme's catalytic activity. Furthermore
      it is quite common especially in older people, to find the im-
      munological activity for a given enzyme remaining relatively
      constant whereas the enzyme activity in the blood falls with
      age. This is thought to be due to the production of increas-
      ing numbers of defective molecules as the individual ages. It
      is necessary therefore to find an approach which makes use
      of the *real* biological activity of enzymes, ie their catalytic
      ability.

2.2   UNITS OF ENZYME ACTIVITY

The need to measure enzyme activities rather than molecular concentrations has an influence on the units in which their measurement is expressed, and it is appropriate to say something about these
at this point. Enzyme studies began many decades ago when laboratories were smaller and communication between them, either directly or via scientific publications, was much less well developed
than it is today. Under these circumstances when individual scientists devised their own methods for measuring enzyme activities
(notice I did not write concentration) it was logical that they would
express their results using units based upon their own particular
methods. This of course led to the development of a variety of units
for each enzyme, and apart from any genuine experimental differences due to the different methods employed, it became progressively more difficult to compare the numerical results from different laboratories without some means of converting them from one
to another.

Thus in one study amylase activity measured by the modern kinetic method yielded a population mean value of 0.124 International Units cm^{-3} with a standard deviation of 0.02 U cm^{-3}; but when measured by the older Somogyi method the population mean was 1.38 Somogyi units cm^{-3} with a standard deviation of 0.31. You can see how it becomes difficult to compare the relative merits of these methods, or to discuss clinical conditions resulting in deviations from the norm, when the results are expressed in such different units and have such different numerical values.

Fig. 2.2a gives a detailed example of the problems involved in the units of measurement of alkaline phosphatase.

The situation was very similar to the use of common regional names for plants and animals which led Linnaeus and others to devise the Latin binomial names as an international standard means of description. To help clarify this situation for enzymes 'The Commission on Enzymes of the International Union of Biochemistry' devised in 1961, the International Unit of Enzyme Activity, orginally abbreviated to IU but now more commonly to U, which for the majority of enzymes is defined as the 'amount of enzyme capable of metabolising 1 μmol of substrate per minute'. This definition would not satisfy more rigorous scientists since it does not specify any aspect of the environmental conditions, temperature, pH etc. Perhaps you might want to consider why this is?

Unfortunately from this point of view, enzymes can vary dramatically in their environmental requirements so a single experimental pH, ionic concentration, etc simply cannot be included in the definition as general for all enzymes. The actual operational conditions are specified in the unit for a given enzyme but may be different for different enzymes and even for a given enzyme when grossly different but popular assay methods are available and commonly used. Thus some difficulties in scientific communication can still occur.

Reaction	The Unit	Unit definition	Originator
phenyl phosphate (PP) → phenol + phosphate	the KA	the amount of enzyme in 100 cm^3 of serum which would liberate 1 mg of phenol from PP in 30 minutes at 37 °C and pH 10	King and Armstrong (1934)
as above	the KK	as above but in 15 minutes	King and Kind (1954)
β-glycerophosphate (BGP) → glycerol + phosphate	the Bodansky	the amount of enzyme in 100 cm^3 of serum which would liberate 1 mg of phosphate from BGP in 60 minutes at 37 °C and pH 8.6	Bodansky (1932–7)
p-nitrophenyl phosphate (PNP) → nitrophenol + phosphate	the BLB	the amount of enzyme capable of hydrolysing 1 mmol of PNP in 60 minutes	Bessey, Lowry and Brock (1946)
as above	the BMC	the amount of enzyme capable of hydrolysing 1 μmol of PNP in 1 minute	Bowers and McComb (1975)
any	the U	equivalent to the BMC	International Unit

Thus 1 Bodansky unit = 0.535 U; 1 KK unit = 0.71 U; 1 KA unit = 1.4 U; 1 BLB unit = 16.7 U; 1 BMC unit = 1.0 U

Notice the variety of substrates used, which indicates the relatively poor substrate specificity of ALP. You can see from the formulae (Fig. 1.3a) that the only common feature of the substrates is a readily accessible phosphate group.

Fig. 2.2a. *Units of measurement for alkaline phosphatase based upon different assay methods*

In addition to the International Unit there is also one that is more compatible with the principles of the SI units: the Katal. This is defined as 'the amount of enzyme metabolising 1 mole of substrate per second' and is unfortunate in being very large. 1 Kat $= 6 \times 10^7$ U and hence 1 U $= 1.667 \times 10^{-8}$ Kat. Its use was supported by the Enzyme Commission in 1972 and it is in regular use in Scandanavia, but elsewhere it has not proved popular.

∏ α-glucosidase is an enzyme which shows only group specificity and a range of substrates can be used for its assay. Use the substrates and conditions listed below to:

(a) suggest the sort of units and expressions of activity level in the sample volumes used in the assays, that the originators of these methods might have devised, and

(b) state the International Unit for the enzyme using p-nitrophenyl-α-glucoside as the substrate.

Incubated for 10 minutes at pH 7, 30 °C using a sample volume of 0.1 cm^3

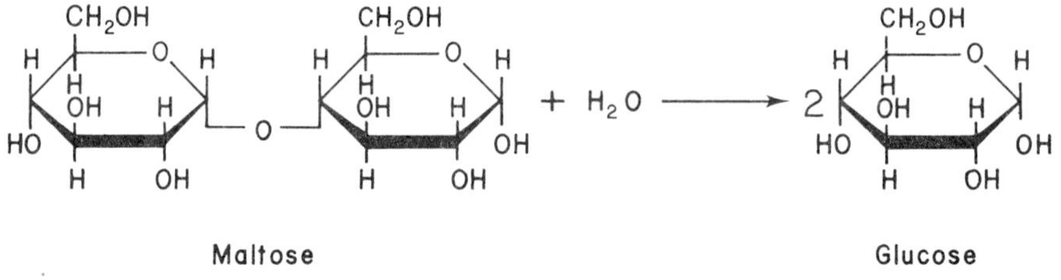

CH2OH
H
O
H
OH
HO
O
OH
+ H2O
2
Maltose
Glucose

Incubated for 5 minutes at pH 6.5, 37 °C using a sample volume of 1.0 cm^{-3}

(*iii*)

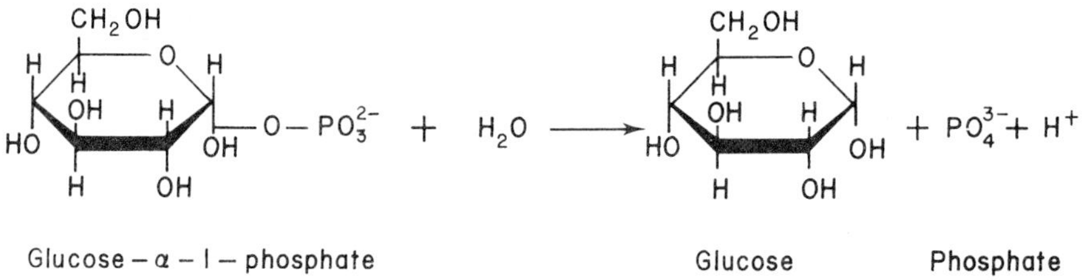

CH2OH
H
O
OH
HO
O – PO3 2–
+ H2O
+ PO4 3– + H+
Glucose – α – 1 – phosphate
Glucose
Phosphate

Incubated for 15 minutes at pH 8.0, 25 °C using a sample volume of 5.0 cm^3

(*a*) Suitable units might be:

(*i*) 'The amount of enzyme in 0.1 cm^3 (although more probably multiplied up to 1 cm^3) of sample capable of producing 1 mg (or 1 μmol) of p-nitrophenyl product or hydrolysing 1 mg (or 1 μmol) of substrate in 10 minutes at pH 7 and 30 °C'.

(*ii*) 'The amount of enzyme in 1 cm^3 of sample capable of producing 1 mg (or 1 μmol) product* or hydrolysing 1 mg (or 1 μmol) substrate in 5 minutes at pH 6.5 and 37 °C'. (*Note the problem here that 1 molecule of substrate will give 2 molecules of a single product and hence the unit values would not be the same).

(*iii*) 'The amount of enzyme in 5 cm^3 of sample capable of producing 1 mg (or 1μmol) phosphate product or hydrolysing 1 mg (or 1 μmol) substrate in 15 minutes at pH 8 and 25 °C'.

(*b*) The proper International Unit would be 'The amount of enzyme hydrolysing 1 μmol of p-nitrophenyl-α-glucoside per minute at pH 7 and 30 °C'. (In practice. for this assay method a division by 10 would be involved to reduce the time to 1 minute and the result would probably be expressed as U cm^{-3} or U dm^{-3}, which would involve an allowance for the 0.1 cm^3 of sample used)

So, to summarise, we usually measure enzyme activity as product accumulated over a selected time (eg μmol min^{-1}) and refer to this as units of activity. If necessary these can be expressed as U dm^{-3} or perhaps as U (mg protein)$^{-1}$ if there is a need to state the amount of enzyme in relation to the total protein in an extract. This latter expression is quite a good way of allowing for experimental variation in the extent of extraction from biological tissue, which is of importance in certain comparative experiments.

2.3. CALCULATION OF ENZYME ACTIVITY

Let us finish off this Part by looking at a few specimen calculations of activity. We will develop this point later but bear in mind for now that the reaction will be followed by measuring the light absorption of the accumulating reaction product, and a typical reaction profile is shown in Fig. 2.3a.

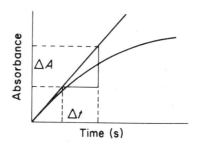

Fig. 2.3a. *The change in absorption by the product during the course of an enzyme reaction*

As this graph curves off quite quickly it is necessary to draw a tangent at the zero time point to obtain a measurement of the so-called 'initial rate of reaction'. From this we can determine the slope and say that this reaction, under *these* conditions produces an absorbance change of ΔA in Δt seconds or

$$\Delta A \frac{60}{\Delta t} \ \text{min}^{-1} \tag{2.3a}$$

We have hit two snags straightaway in that to convert this absorbance change measurement into quantity of substrate actually metabolised by the enzyme (which is required by our definition of International Units), we need to know:

— the relationship between the number of product molecules generated per substrate molecule, and

— the relationship between absorbance and concentration of the product.

This latter relationship is indicated by a physical factor known as the molar extinction coefficient. It is unfortunately the case that this value is not readily available for many compounds, or it may be quoted at a different wavelength or in a different solvent from the ones used in your experiment. Even if this is not the case there can be serious problems over its accurate determination and Varley *et al* (1980) has an interesting discussion of this problem with regard to NAD^+.

In practice therefore most laboratories use a different approach which involves the preparation of a so-called 'calibration graph' which effectively means that the measuring system is calibrated using solutions of the reaction product with a range of known concentrations. It is possible to derive from such a graph a measurement of the amount of product generated by the enzyme in the reaction mixture and hence calculate the enzyme activity. The following example should make this clear.

Let us assume that a series of solutions of the reaction product has been prepared and taken through any chemical or other procedures necessary to produce a measurable derivative and then measured by the technique appropriate to this material. We will assume that the measurement is spectrophotometric (although as we shall see later this is not necessarily the case), and that the Beer–Lambert Law is obeyed. The calibration graph would be similar to the one shown in Fig. 2.3b.

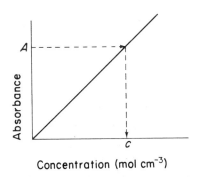

Fig. 2.3b. *A calibration graph to relate absorbance to the concentration of the absorbing compound*

The slope: $$k = A/c \text{ cm}^3 \text{ mol}^{-1} \qquad (2.3b)$$

and hence any concentration c is given by:

$$c = A/k \text{ mol cm}^{-3} \qquad (2.3c)$$

Our enzyme has a reaction profile shown in Fig. 2.3a and the initial rate of reaction can be deduced by taking suitable measurements at convenient points on the zero time tangent. Two time points may be used to determine the slope but the question of how, in practice, this profile is followed in different instruments and in manual investigations is an interesting one that we shall return to in Part 6.

The rate of reaction is given by the rate of change of absorbance

$$= \frac{\Delta A}{\Delta t} \times 60 \text{ min}^{-1} \qquad (2.3a)$$

$$= \frac{\Delta A}{\Delta t} \times 60 \times \frac{1}{k} \text{ mol cm}^{-3} \text{ min}^{-1} \qquad (2.3d)$$

$$= \frac{\Delta A}{\Delta t} \times 60 \times \frac{1}{k} \times 10^6 \text{ } \mu\text{mol cm}^{-3} \text{ min}^{-1} \qquad (2.3e)$$

$$= \text{U of enzyme activity cm}^{-3}$$

(assuming 1 molecule of measured product is generated per molecule of substrate consumed).

The above relationship allows us to calculate enzyme activities cm^{-3} of reaction volume and provided allowance is made for the volume of enzyme sample used, the activity level in the original sample can be determined.

Like many apparently complex calculations the reliability and usefulness of the result depend critically upon the quality of the original data and hence on the experimental design as well as the skill with which it is executed. Scientists are as capable as anyone else of building houses on foundations of sand and in particular the production of a good calibration curve is crucial but not always easy to achieve. Among the problems encountered are

— the production of standard solutions of product with accurately known concentrations;

— imprecision, ie lack of quantitative reproducibility, in the reactions involved in the production of measurable derivatives of the enzyme product;

— difficulty in determining whether the measured product behaves identically in a complex mixture such as blood or urine as it does in the simple solution used to produce the calibration curve;

— the possibility of other components in a complex sample interfering with the derivatizing reaction for the product;

— the possibility of these components reacting independently with the derivatizing reagents.

SAQ 2.3a	You are investigating the activity of an enzyme in a blood sample by using a spectroscopic technique to follow the appearance of reaction product. You have produced the data supplied in Fig. 2.3c (*i*) for a calibration curve to relate absorbance to product concentration.
	In your actual experiment you are able to follow the accumulation of product directly by the absorbance changes produced over a period of time when 100 μl (0.1 cm^3) of sample are used, Fig. 2.3c (*ii*)
	Calculate the enzyme activity cm^{-3} of blood sample. $\longrightarrow$

SAQ 2.3a (cont.)

(i) Absorbance of reaction product		(ii) Changes in absorbance during the reaction	
Concentration (μmol cm^{-3})	Absorbance	Time (s)	Absorbance
10	0.065	2	0.070
30	0.195	4	0.135
50	0.325	6	0.200
70	0.455	8	0.265
90	0.585	10	0.320
110	0.715	12	0.375
		14	0.420
		16	0.460
		18	0.500
		20	0.530
		22	0.560
		24	0.590
		26	0.615
		28	0.630
		30	0.645

Fig. 2.3c

2.4. THE CHOICE BETWEEN MEASUREMENT OF REACTION SUBSTRATE OR PRODUCT

Now let us take the simplest kind of reaction that an enzyme might catalyse, the conversion of a single substrate A to a single product B:

$$A \xrightarrow{\text{enzyme}} B$$

We could follow the reaction, and hence gain an indication of the amount of enzyme activity in a sample, by either measuring the decrease in a concentration of A or the increase in the concentration of B over a period of time. The greater the enzyme activity in the sample the faster these two changes should occur. Notice that time is of some importance here and we shall return to this point later, but remember for now that with regard to enzymes we are interested in *rate* of reaction, in other words the amount of change in a certain period of time. This is because if we simply measured the amount of product generated then eventually a dilute (ie weakly active) sample could give as much product as a more concentrated (ie active) one. It is therefore necessary to state the time interval involved in the measurement.

Another point to note is that the methods available for the measurement of A and B generally determine concentration, ie the quantity in a given volume, and you must bear in mind that for a given quantity the concentration will change depending upon the volume of the assay system in which it is dissolved. While the concentration of materials is often very important for a variety of reasons including its effect on reaction equilibrium, various inhibitory processes, stimulation of alternative reactions in crude mixtures etc, the units of enzyme activity are defined in terms of rates of conversion which involve *quantities* of reaction components and are therefore independent of reaction volume. Bergmeyer (1983) gives a useful discussion of this point.

Π   The concept of concentration is so important and biologists
     so frequently have trouble with this type of simple mathemat-
     ics that it is perhaps worthwhile to carry out a little exercise
     in calculating and interconverting some specimen concentra-
     tion values. If you are unfamiliar, or need refreshment over
     the SI units involved you may find it useful to refer to Table
     2 at the end of this Unit.

     Use the values below to calculate the concentration in μmol
     cm^{-3} in each case.

	Quantity	Volume	M_r of the compound
(i)	15 μmol	3.0 cm^3	–
(ii)	1 μmol	1.0 dm^3	–
(iii)	10 mmol	5.0 cm^3	–
(iv)	6 nmol	0.1 cm^3	–
(v)	75 μg	1.0 cm^3	250
(vi)	232 pg	6.0 cm^3	119

     The answers are as follows:

Concentration (μmol cm^{-3})	Factor		Converts
(i) 5		15/3	3 cm^3 → 1 cm^3
(ii) 10^{-3}		1/10^3	1 dm^3 → 1 cm^3
(iii) 2 × 10^{-3}		10/5	5 cm^3 → 1 cm^3
	then	1/10^3	mmol → μmol
(iv) 6 × 10^{-2}		6 × 10	0.1 cm^3 → 1 cm^3
	then	1/10^3	nmol → μmol
(v) 0.3		75/250	μg → μmol
(vi) 0.325 × 10^{-6}		232/119	pg → pmol
	then	1/10^6	pmol → μmol
	then	1/6	6 cm^3 → 1 cm^3

Returning to our reaction we have said that two approaches to the study of the reaction are possible. Why don't you consider the practical aspects of this for a while and ask yourself if there is any real practical difference between choosing to follow the disappearance of A or the appearance of B.

In fact there is a difference in the two approaches and in general it has been found to be much better to try and follow the appearance of a product in a reaction; the main reasons for this may take a while to explain but are basically straightforward. In most experiments we are interested in comparing enzyme activities, eg values in different extracts, or the same extract under different conditions, or in the presence of different inhibitors etc. It is a fundamental point in experimental design that one tries to have only one variable factor at a time, so that any experimental differences found can be reliably related to that factor. As a consequence of this requirement it is important that conditions are arranged so that the enzymes are able to work at their maximum possible rate under those circumstances. You do not want to be in a position where you are uncertain whether a low activity is due to the inhibitor added as part of your experimental investigation or is due to the experimental conditions being poor and the enzyme working slowly.

Now one aspect of this is that it is usual to supply the enzyme with a significant excess of substrate so that the rate of reaction will not be limited by insufficient substrate molecules as the reaction profile shown in Fig. 2.4a indicates would happen at low (sub-saturation) substrate concentrations.

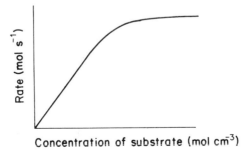

Fig. 2.4a.   *The effect of substrate concentration on the rate of an enzyme-catalysed reaction*

This situation is of course completely different from the one found in the cell where substrate concentrations are generally low but it does create maximum velocity conditions which are more likely to be reproducible. However from the point of view of the present discussion it is much more difficult to measure the *disappearance* of A if the starting concentration is *high*, compared with measuring the *appearance* of B when the starting concentration is *zero* – hence this latter approach is usually taken.

However there are, of course, a number of examples of enzyme measurements for which this is not the case, and which are measured by following changes in substrate concentration. This is usually because they happen to show fairly dramatic changes in absorbance when the substrate is metabolised; a good example is the measurement of uricase by following the fall in absorbance at 293 nm as the enzyme degrades the strongly absorbing unsaturated ring system of uric acid.

Π   Assuming the concentration of substrate supplied is initially high and the enzyme begins to work at maximal rate, draw a graph to show the changes in concentration of product accumulating over a fairly long period of time and another to show the changes in rate of reaction over this time. Briefly explain why you have drawn these profiles.

You should have developed two graphs similar to those in Fig. 2.4b.

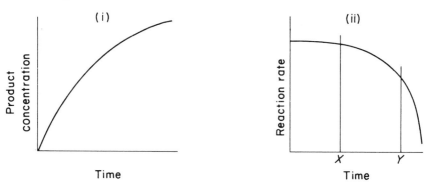

Fig. 2.4b.   *Changes occurring during the course of an enzyme reaction*

The difference between these two graphs arises from the fact that as the enzyme continues to work, more product molecules will accumulate. Eventually of course the system will begin to run short of substrate and the enzyme molecules will have to wait for the increasingly more scarce substrate molecules to diffuse to them, so that the accumulation of product begins to level off (graph i). It will eventually reach a plateau when the reaction reaches equilibrium.

When this pattern of events is displayed as a reaction rate (which is the amount of change per unit time) against time itself (this does make sense!), then in the early stages when all the enzyme molecules can work flat out one has a constant rate of say 10 μmol substrate being converted per second, so the graph is horizontal at that value (graph ii). However, when the substrate begins to become depleted, while the product will not disappear and remains permanently represented on the other graph, the rate falls and will eventually reach zero when the enzyme can no longer find substrate molecules to catalyse, or more likely when the rates of forward and backward reaction are equal.

2.5. THE IMPORTANCE OF THE INITIAL REACTION RATE

Now the changes in profile of these two graphs do create a potential problem in comparative experiments, can you think what this might be?

The problem arises in deciding the point on the graph where you are going to take your measurement for comparative purposes. You could have two enzyme systems in your experiment having exactly the same activity and hence giving you the same profile but if you took time position X for one and Y for the other you could be led as into thinking that there was a real difference between them. The way out of this is to standardise on some time point and in fact the rate at the beginning (called initial or zero time rate) is used because it is obviously the most reproducible and reliable. At this time the enzyme has ample substrate and the chances of the enzyme being denatured or being inhibited are as low as they are ever likely to be.

SAQ 2.5a

The following is a graph illustrating the rate of respiration by a yeast culture.

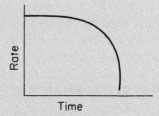

Give three of the possible explanations for the decline in rate with time.

Summary

Some fundamental aspects of the measurement of enzyme activity have been discussed, in particular the concept of the measurement of enzyme activity rather than enzyme molecular concentration, the calculation of enzyme activity and the units in which the activity is expressed. The relative merits of the measurement of disappearing product or appearing substrate, and the importance of the measurement of initial reaction rates are considered.

Objectives

You should now be able to:

- discuss the concept of the measurement of enzymes by determination of biological activity rather than molecular concentration, and show the problems caused by this approach;

- state the units in which enzyme activity is measured and carry out some calculations of enzyme activity from specimen data;

- discuss the merits of the measurement of the appearance of product compared with the disappearance of substrate;

- discuss the necessity for measurement of initial reaction rates as a means of quantifying enzyme activity.

3. Spectroscopic Measurement of Reaction Product

Overview

The commonest approach to the measurement of enzyme activity is to use spectroscopic techniques to measure the accumulation of reaction product. The principles and instrumentation of the main types of spectroscopy are dealt with in other ACOL Units and hence here we shall concentrate on specific procedures used to determine the reaction product directly or indirectly. The discussion will cover the principles, merits and problems of these techniques.

Let us now consider the means by which the reaction product can actually be measured in practice, beginning with the most common approach which is to use spectroscopic methods.

3.1. DIRECT MEASUREMENT

Sometimes the product absorbs uv radiation or visible light of sufficiently different wavelength from that of the substrate to enable us to measure its appearance directly. A method based on this has the merit of being subject to comparatively fewer errors because of its simplicity. Furthermore it also allows us to monitor the reaction very closely by using devices which automatically draw the changing absorbance as a single line on a chart recorder. This has a significant effect on the quality of the results as we shall see in subsequent sections of this Unit.

3.2. PRODUCTION OF DERIVATIVES

However, if direct measurement is not practicable then a common procedure is to form measureable derivatives of the immediate product. This derivatization approach can either be chemical or involve the use of additional (ancillary) enzymes in which case the enzyme is being used as a reagent – an interesting topic dealt with more thoroughly in other ACOL Units.

3.2.1. Chemical Derivatives

An example of a chemical approach is the reaction of the phosphate product, generated by the action of phosphatases on their substrates, with a reagent containing molybdate salts. This produces a strongly absorbing and easily measured blue product called 'molybdenum blue'. A drawback to this method is that the derivatizing reagent frequently inhibits enzyme activity and so the addition of the reagent will stop the reaction. The fact that only one measurement can be taken is a serious problem.

Π   Take as an example an experiment involving the measurement of the activity of an enzyme in which only a single measurement (X) of the absorption of the reaction product was obtained. This value was used to calculate the enzyme on the assumption that the reaction had followed a linear profile as shown in Fig. 3.2a.

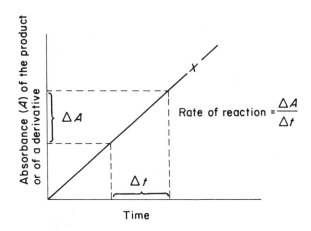

Fig. 3.2a. *Absorbance–time plot for an enzyme reaction*

It is possible that the reaction could have followed a different profile in which case the calculated rate may be incorrect. Fig. 3.2b shows two of these alternative profiles.

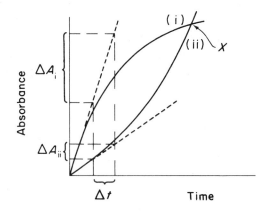

Fig. 3.2b. *Alternative absorbance–time plots for an enzyme reaction*

Note they all pass through point X but produce different absorbance changes over the selected time period represented by Δt and hence in fact have different initial reaction rates. What factors might lead to the production of these profiles? In formulating your response you should consider that biological samples may contain very variable enzyme activities, inhibitory and activating compounds and low concentrations of alternative substrates.

Profile (i) shows a rapid initial rise followed by a progressive decline in reaction rate. It could be due to a number of causes for example:

— the use by the enzyme of substrates present in the sample in addition to that added in the assay reagent;

— a steady decline in rate due to substrate or coenzyme consumption which would be particularly noticeable in a very active specimen;

— the relatively early appearance of denaturation or inhibition effects in an otherwise active specimen.

Profile (ii) shows a lag phase, which is commonly exhibited by enzymes which require activation before reaching full activity. The activation might be by the priming of coupled reactions or the need to regenerate oxidised thiol groups for example.

An alternative approach to the direct addition of derivatizing reagent is to remove samples from the experimental mixture at selected time intervals and add them to the reagent in a separate vessel (the so-called discontinuous approach). This does result in the production of several measurements but even so a poorer line will ultimately be drawn than can be derived from continuous measurements. This is partly due to the errors involved with these additional steps and partly due to having fewer points through which to draw the line.

∏ Plot the data given (Fig. 3.2c) and consider for a moment the problem of drawing the correct line through them.

Parameter X	Parameter Y	Parameter X	Parameter Y
5	6	105	102
10	10	110	106.5
15	16	115	111
20	21	120	116
25	26	125	120
30	31	130	124
35	36	135	129
40	40	140	134
45	45	145	138
50	50	150	142
55	55	155	146
60	59	160	151
65	64	165	155
70	69	170	159
75	74	175	164
80	78.5	180	168
85	83.5	185	171.5
90	88	190	176
95	93	195	180
100	97	200	184

Fig. 3.2c

When plotted the data should produce a very gentle curve (Fig. 3.2d)

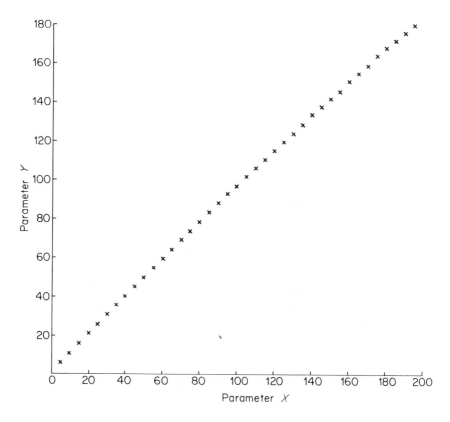

Fig. 3.2d.   *Plot of data of Fig. 3.2c*

There is often great difficulty in knowing whether the data fall on a straight line or on a shallow curve, and this difficulty is increased if relatively few points are available. One trick that is sometimes useful in these situations is to hold the paper at a shallow angle and look along the line when it may become obvious that the data really fit a shallow curve.

An apparently trival but nonetheless true point that should become obvious from these data is that it is always possible to draw a straight line through any set of points! Whether it reflects the correct response and how accurately it does so may be debatable. One common error is to regard the final point or points as more significant than the rest.

One expected advantage of the introduction of computers and micro-processors into analytical laboratories is the general improvement in the quality of line drawing through the experimental data due largely to the lack of bias and reduction in operator error of a properly programmed system.

Manufacturers of the so-called continuous flow automatic analysers have overcome the problem of inhibition by the derivatizing reagent by including a dialyser unit in their apparatus. As the reacting sample flows along the tubing the smaller molecular species can dialyse away from the enzyme as they are produced, into a flow of reagent which will then generate the measurable derivative continuously and without any danger of the reagent inhibiting the enzyme (Fig. 3.2e)

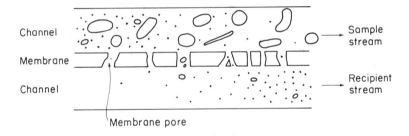

Fig. 3.2e.   *The principles of dialysis*

The very wide range of chemical changes involved in enzyme-catalysed reactions means that an enormous number of chemical derivatization procedures exist and an overall survey is impossible. However a number of reactions have found widespread application and can be examined for the purposes of illustrating the concept.

A suitable example comes from the use of tetrazolium salts in the measurement of oxido-reductase reactions. Well over 350 oxido-reductase enzymes using NAD^+ or $NADP^+$ as their coenzyme are known, many are reasonably non-specific and are capable of reducing other substances including dyes. Over the past decades a number of different dyes have been employed but nowadays members of the tetrazolium group are most commonly used.

These dyes are salts of aromatically substituted tetrazole. Fig. 3.2f shows the fundamental structure of the monotetrazolium group; dimers exist (ditetrazoliums) in which the tetrazole units are linked through additional aromatic ring systems.

$$
\begin{array}{c}
N - NR_2 \\
\| \quad | \\
R_3C \diagdown {}_{N \diagdown} {}_{N^2} {}^{N^+ R_1}
\end{array}
\qquad
\left[
\begin{array}{cc}
N - NR_2 & R_2N - N \\
\| \quad | & | \quad \| \\
R_3C \diagdown {}_{N \diagdown} N^+ - R_x - R_y - {}^+N \diagdown {}_{N} CR_3
\end{array}
\right] 2Cl^-
$$

     (i)   (ii)

Fig. 3.2f. *Basic structure of* (i) *monotetrazolium group* (ii) *ditetrazolium group* (R_1, R_2, R_3, R_x *and* R_y *are aromatic ring systems)*

A wide range of such dyes exists many of which are very useful but cannot be reduced directly by NADH or NADPH and hence some intermediary reducing agent is required; this is usually either a flavoprotein enzyme or an organic compound such as phenazine methosulphate (PMS).

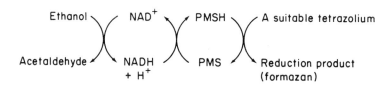

Fig. 3.2g. *Reaction sequence for the formation of a formazan during the reduction of ethanol to acetaldehyde (ethanal)*

Note that for convenience and clarity the abbreviation $NAD(P)^+$ will be used in the text when either or both NAD^+ or $NADP^+$ are relevant to the matter being discussed.

Tetrazolium salts have a number of advantages including a very substantial increase in colour intensity on reduction, the monotetrazolium compounds becoming deep yellow or red and the ditetrazolium compounds becoming deep purple or blue. The reduction is irreversible and the products (collectively called formazans) are non-autoxidisable. Both of these factors make quantitation more reliable.

The molar absorptivities (extinction coefficients) of NADH and NADPH are about 6.3 m^2 mol^{-1} at 340 nm whereas the tetrazolium salts are about 2 to 4 times as high; 2-(p-iodophenyl)-3-(p-nitrophenyl)-5-phenyltetrazolium chloride (INT), having a value of 19.4 m^2 mol^{-1} for example. This difference results quite commonly in a lowering of the detection limits by 2 to 3 fold compared with the direct measurements of NAD(P)H to be discussed later.

Tetrazolium salts do present problems in their use in particular that they are somewhat light sensitive, they are slowly reduced by thiol compounds and more rapidly by vitamin C in the sample. Their reduction products are insoluble and need to be solubilised by detergents but these do however substantially increase the reaction rate. In addition the PMS linking reagent and its reduction product are themselves light sensitive.

However these problems have not substantially curtailed their usefulness and they have proved valuable in perhaps two main areas of application.

— One of these is where a visualisation of location rather than a quantitative measure of activity is the primary requirement. The dyes can be used to indicate the presence and location of oxido-reductase enzymes in tissue sections (a technique known as histochemistry), and on the support medium following enzyme separation by electrophoresis (see the example of lactate dehydrogenase isoenzyme separation discussed in Section 9.1).

— The other is the more typical enzyme assay of $NAD(P)^+$ dependant oxido-reductases and flavin dependant oxidases. They are usually discontinuous systems but when they can be operated continuously they are especially useful if the equilibrium does not favour the product, as the continual removal of enzyme product by the dye substrate will maintain a constant reaction rate for a longer period.

3.2.2. Enzyme Generated Derivatives

While chemical derivatization techniques are commonly employed one major advantage of using enzymes as derivatizing agents is that they are not usually inhibitory to the main reaction. They can therefore be included in the initial reaction mixture to give the desired continuous measurements, and they will also have the advantages associated with the continual removal of product that were mentioned previously.

(a) Oxido-reductases

The object of using ancillary enzyme reagents in this way is to produce a measureable product. While a wide range of such products could in principle be used, in practice most of the coupled enzyme systems actually employed carry out one of a closely related group of changes. The enzymes used for this are properly known as oxido-reductases but are more commonly referred to by more explicit older terms such as reductases and dehydrogenases. They require a coenzyme of the nicotinamide group (ie NAD^+ or $NADP^+$ – or their reduced forms), and may or may not be specific for one or the other. The particular advantage of these enzymes is that the reduced forms of the coenzymes strongly absorb ultraviolet radiation (see p. 57 and Bergmeyer 1983, vol 1 p. 284) and thus the reaction is easily followed by measuring the increase or decrease in A_{340} depending on the direction of reaction. In addition both of these coenzymes are readily available (although $NADP^+$ is rather more expensive especially in its reduced form) and spectrophotometers capable of operating at 340 nm are common and cheap by present-day standards. The reagents are stable; sometimes sufficiently so for them to be amenable to packaging in dried strips for the rapidly developing thin film assay techniques (Steinhausen and Price, 1985). It is necessary to exercise some care since solutions of NADH and NADPH seem to develop enzyme inhibitors on standing but nonetheless the use of ancillary enzymes of the oxido-reductase type is widespread.

ΙΙ   (*a*) Consider the following hypothetical reactions which involve nicotinamide coenzymes, and determine what spectrophotometric changes one could use to measure each of them.

$$(i) \quad X + NADH + H^+ \rightarrow XH_2 + NAD^+$$
$$(ii) \quad YH_2 + NADP^+ \rightarrow Y + NADPH + H^+$$

(*b*) Examine the formulae of glutamate and oxoglutarate given below in a simplified version of a reaction between them.

Deduce the nicotinamide coenzyme change which could occur during the reaction, and which would therefore be used for its measurement.

$$
\begin{array}{ccc}
COO^- & & COO^- \\
| & & | \\
CH.NH_2 & & C = O \\
| & & | \\
CH_2 \quad + \quad H_2O \longrightarrow & CH_2 \quad + \quad NH_3 \\
| & & | \\
CH_2 & & CH_2 \\
| & & | \\
COO^- & & COO^-
\end{array}
$$

Glutamate   2 – Oxoglutarate

(*a*) Both reactions involve the use of NAD type coenzymes and hence spectrophotometric changes at 340 nm would occur. Reaction (*i*) would produce a decrease in absorbance as NADH was oxidised and reaction (*ii*) an increase in absorbance as $NADP^+$ was reduced.

(*b*) This reaction can involve either of the coenzymes depending upon the specificity of the particular enzyme used, but as written it is an oxidation, hence an increase in A_{340} would occur.

$$
\begin{array}{ccc}
\text{COO}^- & & \text{COO}^- \\
| & \text{NAD(P)}^+ & | \\
\text{CH.NH}_2 \quad + \text{ H}_2\text{O} \longrightarrow & & \text{C} = \text{O} \quad + \text{ NH}_3 \\
| & & | \\
\text{CH}_2 & \text{NAD(P)H} & \text{CH}_2 \\
| & + \text{ H}^+ & |
\end{array}
$$

A good example of the use of oxido-reductase derivatizing enzymes is the assay of a clinically important aminotransferase enzyme alanine transaminase (ALT) in which the pyruvate product can be used to generate NAD^+ by employing lactate dehydrogenase (LDH) as a derivatising enzyme (Fig. 3.2h)

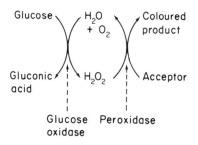

Fig 3.2h.   *Scheme for the assay of alanine transaminase*

In this case the decrease in absorbance at 340 nm as NADH is oxidised reflects the amount of pyruvate available to the LDH enzyme which in turn depends upon the activity of the ALT in the original sample since this actually generates it. In practice the experimental reagent contains 2-oxoglutarate, alanine, NADH and LDH so that the first and second reactions can proceed.

It would be possible to measure the activity of ALT by a chemical assay of one of the reagents involved and indeed methods based upon the reaction of the keto acid pyruvate with dinitrophenylhydrazine (DNPH) are available. Such methods suffer from the problems inherent in chemical derivatization techniques which were described earlier, and in this case a lack of specificity also occurs since the reagent reacts with several similar components in the sample and assay system. The enzyme derivatization approach shown above is therefore a better approach and also illustrates two important general features of the technique namely a substantial increase in specificity together with a decrease in detection limit.

Enzyme nomenclature is a complex business and, as with most other aspects of science, individual components (in this case the enzymes) are given formal, highly descriptive and specific names. They are unfortunately rather long and hence trivial, often old-fashioned names or even lettered abbreviations, are frequently used. To try and prevent undue increase in length of text we shall resort to the latter device after the first occasion on which an enzyme is encountered. Table 5 (Appendix) gives a full description of the enzymes referred to in this Unit.

(b) Hydrogen Peroxide

Several alternative coupling systems which result in straightforward spectroscopic assays are available and among these systems those involving H_2O_2 are quite common and valuable. This is partly because a number of important oxido-reductase enzymes using oxygen as an electron acceptor generate H_2O_2, and also because H_2O_2 is easily measured with a detection limit at least an order of magnitude lower than the direct spectroscopic measurement of NAD(P)H.

Let us examine for a moment the very popular method for glucose assay which employs glucose oxidase to generate glucuronic acid and hydrogen peroxide. Assay of the former is feasible but requires complex chemistry while the latter (H_2O_2) can be assayed in a number of ways including the use of selective electrodes. However it is common to use the enzyme peroxidase to degrade the H_2O_2 as a means of oxidising a suitable acceptor (Fig. 3.2i).

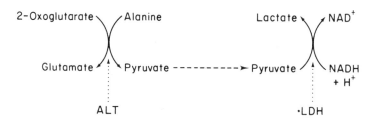

Fig. 3.2i. *Scheme for the enzymic assay of glucose*

In this example the H_2O_2 was produced directly by the enzyme but in other cases it is produced as part of a derivatization process:

$$\text{Phosphatidyl-choline} + 2\,H_2O \xrightarrow{\text{phospholipase}} \text{phosphatidic acid} + \text{choline}$$

$$\text{choline} + 2\,O_2 + \xrightarrow[\text{oxidase}]{\text{choline}} \text{betaine} + 2\,H_2O$$

Spectroscopic systems for the detection of H_2O_2 are of 3 main types.

(*i*)   those involving peroxidase-catalysed oxidation of single chromogens, previously benzidine derivatives such as dianisidine or toluidine but nowadays rather safer alternatives.

(*ii*)   those involving peroxidase-catalysed oxidation of two chromogens of which the very important and versatile Trinder reaction is an excellent example. In the original form this involved the peroxidase-catalysed oxidation of phenol by the H_2O_2, with the oxidation product coupling to aminophenol to give a measureable red derivative. An advantage of the reaction is that the specificity to phenol is low so that the method can be used for the detection and measurement of a number of other phenolic reaction products. Additional advantages are that the reactants and products are highly soluble, non-autoxidisable and stable, and the reaction is rapid with a long linear region to the calibration graph.

(*iii*) of lesser importance is a catalase dependent reaction, in which the catalase acts as a highly specific methanol degrading peroxidase. The formaldehyde (methanal), produced can be used in the non-enzymic Hantzch reaction as shown below. However the high temperature required for colour development prevents this being used as a continuous assay.

$$H_2O_2 + CH_3OH \xrightarrow{\text{catalase}} HCHO + 2H_2O$$

$$HCHO + \text{acetylacetone} + NH_3 \xrightarrow{\text{non-enzymic}} \text{coloured product}$$

One point which may have confused you and is worth clarifying at this junction concerns an aspect of enzyme nomenclature. It is normal practice to name all enzymes of a particular type as if their reactions all proceed best in the same direction, ie all as reductions, carboxylations, deaminations etc. So we have the apparent anomaly above of an enzyme called lactate dehydrogenase being used in the reverse direction to *produce* lactate, which in fact it can easily do because in this case the reaction equilibrium is such that without having to introduce excessively high substrate concentrations the reaction can proceed tolerably well in either direction.

Occasionally these derivatization systems can become quite complex as shown by the assay for the enzyme creatine kinase, CK (or for its metabolite creatine phosphate), in which two sequentially linked derivatizing reactions are used.

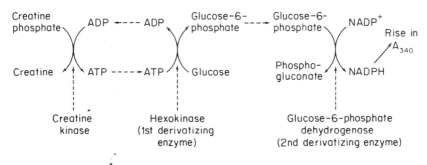

Fig. 3.2j. *Scheme for the assay of creatine kinase*

Now there are some problems with enzyme based derivatization systems. Let us examine them by going back to the ALT assay described earlier (Fig. 3.2h). It is assumed that the decrease in absorbance at 340 nm reflects the activity of the ALT in the sample but this will depend upon no factor being 'rate limiting'. Obviously if insufficient alanine is included then the ALT would be working slowly and a falsely low value would be seen; a similar effect would be observed if too little NADH is included. But this idea applies also to the derivatizing enzyme which it is generally considered should have 10 times the activity of the first enzyme. This avoids rate limitation effects should it suffer partial denaturation or inhibition, or conversely should a sample appear which has substantially higher than normal activity.

∏   What do you think are the important aspects of this requirement for the exclusion of rate limiting factors for a busy routine hospital laboratory analysing many samples a week and needing reliable results for patient diagnosis? It might help to consider our creatine kinase example in this context.

Many laboratories would be concerned about the twin points of possible sources of error due to the large number of reagents and the potentially sensitive steps involved, and also about the possible high cost. In this example the second derivatising enzyme needs to be present at 100 times the activity of the CK in the sample in order to avoid rate limitation effects! It is not surprising that such expensive and error-prone chains are uncommon in routine work.

Another more subtle problem is that coupled systems of this type frequently exhibit a lag phase in their response while the product of the first reaction accumulates and stimulates the second, which of course is the one which provides the measured product. This awkward occurrence can be minimised but not eliminated by the inclusion of the 10-fold excess of ancillary enzyme.

(*c*) *Luminometry*

It should be obvious by this time that notwithstanding the problems discussed above the derivatization technique is a very versatile and valuable one, but it does have its limitations. For reasons of pure chemistry it may not in fact be possible to carry out a single or even a multiple enzyme link to an oxido-reductase, in order to make it possible to measure the enzyme by changes in absorbance at 340 nm. The carboxylase group of enzymes provides a good example of this, since carboxylation and decarboxylation reactions do not involve oxidation or reduction. The technique of luminometry is one that is rapidly assuming importance in clinical chemistry laboratories for a variety of reasons (see other units) including a potential value in overcoming the problem mentioned above.

The term luminometry implies the production of light and a number of biological systems ranging from fungi, deep sea fish, molluscs, cephalopods, various species of plankton and of course fireflies are well known for their ability to emit light. They do this by using the energy of ATP to activate a substrate molecule which on its decay re-emits that energy as visible light. It has proved possible to isolate some of the chemical components involved and produce a system capable of assaying ATP by measuring the light emitted during the course of the reaction. Radioisotope scintillation counters, modified spectrophotometers or more commonly nowadays, purpose-built luminometers are used for the measurement.

Luminometry has a number of distinct advantages as a general technique including a linear response over a wide range of concentrations, down in fact to much lower ones than can be achieved using normal spectrophotometric methods. Detection limits for ATP can be as low as 10^{-13} mol dm^{-3} for luminometry but only 10^{-5} mol dm^{-3} for $NAD(P)^+$ assayed by normal spectrophotometry. Perhaps more important in the present context is that the technique gives a completely different approach to enzyme assay which can be used if coupling to a dehydrogenase or reductase is difficult or presents particular problems.

∏   Again the CK assay (Fig. 3.2j) provides a useful example
       here – can you see how luminometry could be employed to
       decrease the complexity of this assay?

       Without considering the detailed procedures involved one
       could in principle use the technique to follow the appearance
       of ATP as a result of CK activity. This has the advantage
       of removing the need for the 2 derivatizing enzymes (Fig.
       3.2j), although they would have to be replaced by the rather
       simpler single enzyme system necessary for the luminometric
       assay.

The main enzyme system used for the assay of ATP is the luciferase
system of fireflies in which luciferin is oxidised by molecular oxygen:

$ATP + O_2 +$ luciferin $\rightarrow$
$\qquad$ oxyluciferin $+ AMP + PP_i + CO_2 +$ light (562 nm)

After some effort conditions for the production of a constant light
signal with an intensity proportional to ATP concentration have
been developed. Continuous monitoring of the reaction is now
therefore possible. Assays for the production or consumption of
ATP are available although the latter are awkward in attempting
to measure the decrease in concentration of a high concentration
metabolite.

Assays of H_2O_2 are possible using the luminol system shown in the
following equation. The method is useful despite being a discontinuous one.

$$\text{luminol} + H_2O_2 \xrightarrow{\substack{\text{horseradish} \\ \text{peroxidase}}}$$

$$\text{aminophthalic} + 2\,H^+ + 2\,H_2O + N_2 + \text{light (425 nm)}$$
$$\text{acid}$$

Assays of NAD(P)H are also possible using bacterial enzymes especially those isolated from *Beneckea harveyi* or *Vibrio fischeri*

$$NAD(P)H + H^+ + FMN \xrightarrow{\text{diaphorase}} NAD(P)^+ + FMNH_2$$

$$FMNH_2 + RCHO + O_2 \xrightarrow{\text{monoxygenase}}$$

$$FMN + H_2O + RCOOH + \text{light (493 nm)}$$

A number of problems are encountered with the luminescent approach including light quenching effects similar to those found in radioisotope scintillation counting, background fluorescence and luminescence by the cuvette and reagent or sample components. Enzyme inhibition by ions and changes of emission wavelength by pH, temperature and ions are common. However these problems are being resolved and the increasing availability of purpose-built instruments makes this a technique of considerable future potential.

(d) Amplification by Enzyme Coupling

Coupled enzyme derivatization systems have an additional value in that, if necessary, they can be constructed to provide a signal amplification system which is useful if the enzyme activity is very low or the reaction equilibrium strongly against the direction of interest. Most human specimens are likely to have reasonable enzyme activities but particularly in studies of inherited diseases, very low or even near zero activities may occur. For this and a variety of other reasons it is becoming increasingly important to miniaturise assay systems either to reduce costs or to make use of samples which are available only in small quantities. Intra-uterine fetal specimens, extracts from cultured cells, low volume blood specimens from obese, very young or geriatric patients are among these.

Let us consider the case of an enzyme Ex which is producing NADH at a very slow rate. By including two enzymes (and the appropriate substrates, cofactors etc) we can create a cyclic reaction in which the NADH molecules are re-used many times and ultimately produce quite large quantities of a measureable product.

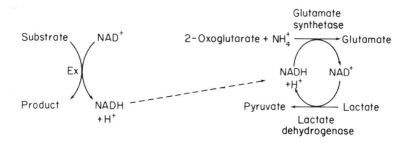

Fig. 3.2k.   *Scheme for amplification by enzyme coupling*

In this example the 2 enzymes glutamate synthetase and lactate dehydrogenase (which incidentally is now working in the direction implied by its trivial name!) can, under suitable experimental conditions, recycle the NADH to produce over 10^4 pyruvate molecules per minute for each NADH molecule produced by the enzyme Ex. This will rapidly generate sufficient pyruvate to produce a useful measurement.

An alternative and simpler means of amplification would be to measure the fluorescence of the NADH but biological samples are complex mixtures which often have many fluorescing compounds in them giving a high interfering background fluorescence.

3.3.   THE USE OF ARTIFICIAL SUBSTRATES

A final spectrometric approach to enzyme assay which is suitable for enzymes of less than absolute specificity is to use an artificial substrate which of course is carefully selected to provide an easily measured product. How far it is possible to move away from the normal substrate for the enzyme depends upon its degree of specificity but in some cases a completely unnatural substrate can be used. α-glucosidase provides a reasonable example as it is an enzyme which only requires an accessible α-glucosidic linkage for reaction:

4-nitrophenyl-α-D-glucoside $+$ H_2O $\longrightarrow$ 4-nitrophenol $+$ glucose

The advantages of the substrate used in the above reaction is that at pH values above 11 the 4-nitrophenol product is bright yellow and hence is easily measured at 400 nm.

Techniques using artificial substrates are exceedingly common because of their availability and ease of use, but it must be remembered that the measured activity is not the true biological activity. This might be of importance to some biologists, although it rarely is in medical laboratory assays where we are usually making comparisons using the same technique between samples from different individuals. A potentially more significant problem however is that the substrate might be used by a range of other low specificity enzymes, which could give variable results due to changes in these other enzyme activities independently of the one of interest.

∏   Trypsin is an enzyme which hydrolyses peptide bonds in protein and peptide molecules. It can be assayed using the artificial substrate benzoyl-arginine-4-nitroaniline.

Can you deduce which part of the substrate might, on hydrolysis of the peptide bond, undergo sufficient changes in structure or properties to produce a measurable change?

The answer is the 4-nitroaniline part which on liberation absorbs 405 nm radiation in contrast to the 315 nm absorption of the original substrate. A continuous monitoring of the reaction at 405 nm is therefore possible. Furthermore it is possible to move the product absorption to a still higher wavelength and increase the intensity of absorption by diazonium coupling of the nitroaniline.

A closer approach to a more natural substrate is possible in this case by using indoxyl ester derivatives of natural peptides. Hydrolysis yields the indoxyl which spontaneously oxidises to indigo type dyes.

SAQ 3.3a

For each of the following give:

(*i*)   a reason why direct measurement of the product of an enzyme reaction might not be possible;

(*ii*)   a potential disadvantage of the measurement of enzyme reaction product by chemical derivatization;

(*iii*) an advantage of the enzyme derivatization technique compared with chemical derivatization techniques;

(*iv*)   a disadvantage of those enzyme derivatisation techniques that involve more than one derivatising enzyme;

(*v*)   an advantage of luminometry as a measuring technique for derivatives of reaction products;

(*vi*)   a disadvantage of using artificial substrates in measurements of enzyme reactions.

Summary

This part discusses the use of spectroscopy in the measurement of the appearance of the reaction product. In particular consideration is given to:

— direct measurements of appearing product, or a suitable derivative of it;

— chemical and enzymic derivatization;

— the amplification of the reaction signal by enzyme cycling of a reaction product;

— the use of artificial substrates;

— measurement by luminometry.

Objectives

You should now be able to:

● explain the advantages of the direct measurement of enzyme reaction product by measurement of its absorption of visible light or ultra-violet radiation;

● explain the discontinuous measurement of enzyme reactions and explain why a single measurement of reaction might lead to errors in the determination of initial reaction rate;

● describe and compare the advantages and problems of enzyme derivatization and chemical derivatization as techniques for the measurement of enzyme reactions;

● state the major advantages of enzyme derivatization compared with chemical derivatization in enzyme assay;

- briefly describe the use of oxido-reductases and hydrogen peroxide generating systems in enzyme derivatisation methods;

- summarise the advantages of luminometry in enzyme derivatization systems;

- briefly describe, or show diagrammatically, the use of derivatizing enzymes as cycling systems for signal amplification in the analysis of samples of low enzyme activity;

- state the advantages and problems of using artificial substrates in enzyme assay.

4. Non-spectroscopic Measurement of Reaction Product

Overview

For some enzyme catalysed reactions a non-spectroscopic measuring technique is more suitable than a spectroscopic one. In this part of the Diagnostic Enzymology Unit some situations in which this occurs are considered, together with some of the most common alternative techniques. The principles of these techniques are covered elsewhere, hence here we shall concentrate on their merits and problems, and on some specific applications.

It is undoubtedly the case that the vast majority of enzyme reactions are monitored by following the appearance of product using one of the spectroscopic techniques just described. However in a number of cases alternative techniques have been employed. Why not pause for a moment to consider possible reasons for this?

Perhaps the most obvious general situation is where the chemistry of the reaction or the nature of the product is such that even with the inclusion of luminometry and fluorimetry, none of the spectroscopic approaches yield an acceptable measurement. Carboxylation and decarboxylation reactions are commonly like this. There is also a reasonable number of occasions when a spectroscopic approach is feasible but maybe time consuming, expensive due to reagent costs, prone to error due to its complexity, use of unstable reagents or for other reasons. Sample preparation can sometimes be excessively involved, for example when it proves necessary to reduce background fluorescence of blood specimens. Samples which are turbid or coloured due to released haemoglobin, or high lipid or bilirubin content, present difficulties in spectroscopic methods due to their high background absorbance. Finally of course a few reactions happen to be particularly amenable to measurement by an alternative technique.

4.1. ELECTRICAL TECHNIQUES

A number of techniques are based upon changes in charged states within the experimental system as a consequence of the reaction. A useful discussion of their relevance to enzyme assay is given in Bergmeyer (1983). Brief discussions of the techniques themselves are given in Holme and Peck (1983), Skoog and West (1985) and Williams and Goulding (1986). Excellent reviews of modern developments in this field and their applications to medical laboratory work are given in Vadgama and Davis (1985), and Siggaard–Andersen (1986).

4.1.1. Conductance Measurements

If a solution contains ions then it should be obvious from basic physics that the solution ought to conduct electricity as the ions migrate under the influence of the applied potential. In general the greater the number and charge of the ions the greater the conductance (sometimes incorrectly called conductivity). How do you think this could be used to measure enzyme reactions?

If the reaction produces a significant change in ion numbers or charge then a change in electrical conductance will occur; this can be used as a measure of enzyme activity after suitable calibration.

However a number of factors mean that equipment design and use for this technique are not as straightforward as might appear at first sight.

— Conductance will depend upon electrical field strength and to maintain this at a constant level electrode size and separation must be fixed which is easy enough. However the potential must also be constant which is less easy to ensure although it is achieved in modern instruments.

— Conductance will also be affected by ionic concentration, and with increasing molecular concentration the degree of dissociation falls, particularly for weak electrolytes, (Ostwalds Dilution Law). Furthermore the ions can interfere with each other and reduce the overall rate of movement to the electrodes (and hence current) if their concentration is high enough.

— Environmental temperature has a marked effect on several aspects of the system leading to a requirement for temperature stabilisation to ± 0.5 K. However this is achieved by modern instruments.

— Enzymic systems are usually buffered for reasons discussed previously (Part 1) and an effort must be made to keep the conductance of the buffer low by using complex organic types (eg TRIS), if the background conductance of the experimental system is not to be excessively high when compared with the anticipated changes due to the reaction. A related problem is that the experimental sample or reaction components will introduce ions into the system, again raising the background level. However most of the commercial instruments are sufficiently sensitive to allow the use of extensively diluted samples.

— On occasions the results found are not those one would predict due to further reactions of the substrates or products with other ionic species within the experimental systems, eg the buffer ions. Life is particularly difficult if such interactions are not constant and vary with the buffer type and concentration, sample volume and composition etc.

∏   Two applications of this technique are shown below. Can you
       very briefly explain the basis of the assay in each case?

       (*i*)   Hydrolysis of acetylcholine by cholinesterase

$$CH_3COO(CH_2)_2N^+(CH_3)_3 + H_2O \xrightarrow{\text{cholinesterase}}$$
       acetylcholine
$$CH_3COO^- + HO(CH_2)_2 N^+(CH_3)_3 + H^+$$
                (acetate)   (cholate)

       (*ii*)   Hydrolysis of urea by urease,

$$CO(HN_2)_2 + H_2O \xrightarrow{\text{urease}} 2NH_4^+ + CO_3^{2-}$$

       (*i*)   The reaction begins with a single ionic species (acetyl-
                choline), and apparently generates 3 others (acetate,
                choline and a proton). Although the proton associates
                with a buffer anion and effectively disappears the in-
                crease to two ions is sufficient to yield a measureable
                response.

       (*ii*)   A really useful measurement is obtainable here since an
                uncharged substrate molecule is converted into three
                charged product molecules, and several commercial in-
                struments use this principle for the measurement of this
                important material (urea). The Beckman BUN Anal-
                yser is one example.

4.1.2. Polarographic Measurements

A different set of physical principles and instruments, usually involving a gas permeable membrane to protect the electrodes and
eliminate reactions due to non-gaseous material, allow the measurement of certain gases. This alternative approach to reaction measurement is termed polarography.

Perhaps the best and certainly among the most common example of
the use of these polarographic systems in medical laboratory work
is the enzyme based measurement of glucose. Glucose can be mea-

sured in a number of ways (Fig. 4.1a), including direct chemical reactions, eg with Benedict's reagent (i), or by the use of glucose oxidase and a derivatizing system involving peroxidase (ii). Both of these yield spectrophotometrically measurable products but both, and especially the former, are subject to interference by other materials, lack of specificity etc.

The use of a polarographic system (iii) avoids the need to include the troublesome peroxidase enzyme, or carry out the even more error-prone chemical assays. In biochemical laboratories the Clark oxygen electrode is perhaps the most commonly used and several successful commercial instruments (eg those produced by Beckmann Ltd and The Yellow Springs Instrument Company Inc) use the very similar H_2O_2 sensitive electrodes for glucose analysis.

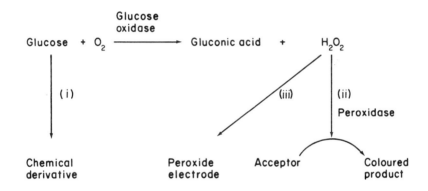

Fig. 4.1a. *Alternative schemes for the assay of glucose*

4.1.3. Potentiometric Measurements

In very simple terms ions produce electrochemical potentials at various boundaries (eg electrodes) in solutions. Since these potentials are concentration-dependent they can be used to measure ion concentration (as in pH or ion selective electrodes) and hence to follow chemical and thus enzymic reactions.

If a metal rod is immersed in a solution of its own ions an electrochemical potential difference (the Galvani potential) arises at the rod surface as metal ions form and pass into solution and then become re-deposited. The potential difference is dependent upon temperature, type of metal and a number of other factors including ion concentration (or more strictly activity). Suitable instrumental systems for the measurement of the emf of a cell comprising an indicating and an appropriate reference electrode are available.

Undoubtedly the commonest application of the principles of potentiometry is in the use of the pH electrode, but relatively few enzyme reactions are monitored by following pH changes occurring as a consequence of the reaction.

∏   Perhaps you might like to consider what is the main reason for this last approach being unsatisfactory. Some of the information provided earlier on the fundamental properties of enzymes and on the ideal profile of the reaction rate might help you here.

The major problem is the effect of pH on the activity of enzymes. Whereas extreme pH values will result in permanent inhibition of the enzyme through denaturation effects, even slight changes away from the pH optimum will reduce the activity of many enzymes. Obviously if the pH is allowed to change during the reaction a progressive decline in reaction rate is likely to occur for this reason in addition to the others described earlier. Determination of a true reaction rate could become quite difficult. If the solution is buffered in order to overcome this problem any protons released during the enzyme reaction will be immediately taken up by the buffer anion, and there will be no detectable change in pH.

To avoid the adverse effects of pH an ingenious approach using an automatic titrator ('pH-stat') is sometimes used. In these systems the pH signal is kept as constant as possible by the continual addition of acid or base, the quantity required in a given time being an indication of reaction rate. Gastric enzymes such as lipases are sometimes measured in this way:

$$\text{triacylglycerols} + H_2O \xrightarrow{\text{lipase}}$$
$$\text{diacylglycerols} + \text{monoacylglycerols} + (\text{fatty acids})^- + H^+$$

Modified pH electrodes (membrane electrodes) located in gas tight chambers have been used to monitor reactions yielding low RMM volatile products with pH properties, eg CO_2 or NH_3

$$\text{amino acids} \xrightarrow{\text{decarboxylase}} \text{amines} + CO_2$$

4.1.4. Thermal Measurements (Microcalorimetry)

For an enzyme reaction taking place at constant pressure the change in enthalpy (heat content) of the whole system is equal to the change in internal energy plus any work done by the system. Since generally speaking little or no work is done in most biochemical assay systems, a change in enthalpy will occur and this will result in a temperature change (positive for endothermic reactions and negative for exothermic reactions). Thus the activity of the enzyme urease at pH 7 can produce the release of 6.5 kJ mol^{-1} of heat. Changes such as these have led to the development and use of the technique of microcalorimetry in selected enzyme assays.

The technique, though lacking intrinsic specificity and hence relying on the use of specific enzymes and pure reagents to minimise side and alternative reactions, should in theory be applicable to virtually every reaction and would therefore have considerable merit. Its suitability for use with unclear specimens (cells, extracts, lipaemic and haemolytic sera etc) which would need extensive and error-prone processing before spectroscopic analysis, is another significant advantage. The biggest problem which severely limited the technique for many years was the small size of the temperature change for the likely enzyme activities and most probable sample volumes. Temperature stabilisation of a very high quality is required if the devices are to be successful but this is now possible in modern instruments. The most significant problems limiting their use in medical laboratories at the present time are the cost of the apparatus, unfamilarity of the technique and perhaps most important the relatively slow response time for the system. This latter point will extend the analysis time and it is not compatible with the needs of medical laboratories to process large numbers of samples and produce the results as quickly as possible.

4.2.   MISCELLANEOUS TECHNIQUES

A wide range of additional methods has been used for particular applications.

— The measurement of large gas volume changes by manometry. This technique was used extensively by Krebs and others when they elucidated the metabolic pathways of respiration in the days before the techniques of chromatography and radioisotope labelling were available.

— The technique of polarimetry in which the changing optical rotation of the substrate as the enzyme metabolises only one of the stereoisomers is followed.

— Direct and detailed chemical investigations of the product. These may involve the use of chromatographic, precipitation, solvent extraction or other techniques for the separation of product from the other experimental components followed by product identification using standards, R_f measurements, specific chemical reaction, or even more sophisticated spectroscopic techniques such as nmr, ir etc.

— In some cases the use of a radioactively labelled substrate is necessary to lower the detection limits dramatically if the enzyme has a fundamentally low activity or is present in small quantities, eg in tissue extracts.

Π   The enzyme glutamic acid decarboxylase will decarboxylate its substrate thus:

$$
\begin{array}{ccc}
\text{COOH} & & \text{NH}_2 \\
| & \text{Decarboxylase} & | \\
\text{CH.NH}_2 & \xrightarrow{\hspace{2cm}} & \text{CH}_2 \quad + \quad \text{CO}_2 \\
| & & | \\
\text{(CH}_2)_2 & & \text{(CH}_2)_2 \\
| & & | \\
\text{COOH} & & \text{COOH} \\
\end{array}
$$

Glutamic acid   4 – Aminobutyric
                                      acid

(*a*) Briefly describe how this reaction might be investigated and measured using the general approaches mentioned in Parts 3 and 4.

(*b*) Perhaps you have noticed that the enzyme could possibly have decarboxylated the substrate via the other carboxyl group. How might you use radioisotope labelling to show this was not the case?

(*a*) It might be possible to measure the aminobutyrate product by a suitable colour reaction if the reaction was sufficiently specific, or following chromatographic separation if the reaction was insufficiently specific. Alternatively, if the substrate was made radioactive eg by labelling with ^{14}C, then radioactive CO_2 could be trapped in an alkaline hydroxide or hyamine solution and the enzyme activity determined by the radioactivity accumulated in the reaction product.

(*b*) If selectively labelled glutamate substrates were used the specificity for carboxyl group could be determined. Thus C_1 labelled glutamate would yield labelled CO_2, whereas C_4 labelled glutamate would not. This could be checked by monitoring the activity of the other product in each case.

In general these approaches are rarely of importance in medical laboratory work since they frequently require high enzyme activities, long reaction times or detailed, labour intensive and time consuming analyses. In short they are not suitable for a routine hospital laboratory situation where high sample throughput, rapid output of results, and maximum result quality with minimal staff skill are major considerations.

SAQ 4.2a

The list below contains a series of definitions of some measuring techniques which can be used to follow changes produced by the consumption of substrate or generation of product during the course of an enzyme reaction. The names of the techniques are also given in a second list, and you should match each definition with it's appropriate name.

The Definitions

(*i*)   The generation of an emf in a cell containing indicating and reference electrodes, by ions in the solution.

(*ii*)   The collection of gases produced by the reaction and the measurement of gas volume.

(*iii*) The change in resistance of a solution during an enzyme reaction.

(*iv*)   The measurement of the change in optical rotation of a solution during the course of an enzymic reaction.

(*v*)   The measurement of the temperature changes occuring during the course of a reaction.

The Names

manometry
calorimetry
polarography
polarimetry
potentiometric measurement
conductance measurement.

SAQ 4.2a

SAQ 4.2b　State one problem associated with the measurement of enzyme reactions using each of the following techniques:

(*i*)　microcalorimetry;
(*ii*)　manometry;
(*iii*) conductance measurements.

Summary

Non-spectroscopic techniques used for the measurement of the appearance of reaction product are considered. In particular the principles and problems of potentiometric, conductance, polarographic and thermal techniques are described. There is a brief survey of other techniques of lesser importance.

Objectives

You should now be able to:

- explain and distinguish the principles of potentiometric, conductance and polarographic techniques and list problems associated with each when used to monitor enzyme reactions;

- give an account of thermal methods of following enzyme reactions;

- discuss, with examples, one other technique to measure an enzyme reaction and explain why this technique is not commonly used in hospital laboratories.

5. Some Problems Associated with the Samples Used for Enzyme Assay

Overview

This part describes the various types of sample that can be taken from humans for analysis. The relative usefulness of these samples is discussed with particular emphasis on blood specimens as the most commonly used clinical sample. The effects of various physiological factors and a number of post-sampling events on enzyme activity are described.

The value of laboratory assays in medicine is critically dependent on the nature of the sample being analysed. It is necessary to ensure that the sample is truly representative of the material in the body, and that it is uncontaminated and not otherwise affected by the sampling process. It must also be unchanged by its transport, storage and subsequent laboratory treatment. These matters are so vital in medical laboratory work that they are dealt with extensively in another unit. However there are some special aspects of this problem with regard to enzyme assay that are developed here. The discussion in Lodensen (1980) will provide additional information.

The Main Types of Sample Used for Enzyme Analyses

Undoubtedly the majority of assays are done on blood, generally for soluble or suspended enzymes present in plasma, although in some cases, for example in the study of inherited diseases, enzymes in red or white blood cells are investigated.

Very few enzymes investigations are undertaken on other biological materials (urine, faeces, sweat, cerebrospinal fluid etc) so only a brief discussion of these subjects will be given at the end of this Part.

5.1. BLOOD SAMPLES

Restricting ourselves to blood then the size, the relative instability and the ease of interference of enzymes raise some problems in the removal, preservation and storage of blood samples. Another problem is that some physiological situations create difficulties in the evaluation of the data obtained for the enzyme assays. The procedures used in attempting to solve these problems are not always compatible with the requirements for the assay of other components in the specimens.

5.1.1. Concurrent Physiological Events

Blood enzyme levels can be influenced by a wide range of physiological changes occurring prior to, or in parallel with, the blood sampling process, and many of these effects take hours or even days to return to baseline. For enzymes present in high activity in cells, it is likely that cellular damage will cause the release of significant amounts of enzyme. AST is such an enzyme and activity rises of 10–90 % are seen following exercise even in trained athletes. Intravenous and intramuscular injections also cause a substantial rise, the extent of which depends upon the material injected as well as the site of injection and the volume used, suggesting that the effect is more complex than one of simple physical damage. In all these cases 3–5 days are required before the AST value will return to the normal level for that individual.

5.1.2. Types of Blood Sample

The vast majority of enzyme assays are done on venous blood samples taken by traditional hypodermic syringe or the more recently developed vacuum devices. However in a number of cases (red) capillary blood is used, generally because access to veins is difficult or the volume available from the veins is low. Among these situations are those occasions when the patients are grossly obese, severely burnt, or are fetuses, babies or geriatrics with fragile veins. It is of considerable importance to know if the enzyme activities are the same in these samples as in the more common venous ones in order that a more reliable judgement can be made as to the likelihood of an abnormality. It is also essential to know if the sample stability, enzyme nature and assay procedure are the same for this type of sample. For many non-enzymic materials (eg glucose, calcium, potassium and total proteins) such venous–capillary differences are well known but for enzymes the information is less certain. Enzymes such as LDH and aldolase seem to show no differences in nature or activity but others with a high cytoplasmic cellular activity can show substantial differences. One study reported that capillary AST levels can be up to 26% higher compared with those in venous plasma. There are many reports of the release of cellular acid phosphatase (AP) isoenzymes with a different isoenzyme composition. Nonetheless the use of such samples is probably increasing in medical laboratory work and the investigations will become more reliable and useful as these variable factors become better understood.

Investigations on arterial blood are uncommon due to the difficulty and danger inherent in obtaining such samples. This is perhaps fortunate since several parameters are significantly different in blood from this source.

Π   Bearing in mind a very simple picture of the transport of ma-
    terials around the body in the blood system (and you might
    like to refer to a simple textbook of biology if you need re-
    freshment on this), can you name any blood component that
    might show a significant difference between arterial and ve-
    nous samples.

In very simple terms the function of the arterial blood supply is to deliver food materials and oxygen to the tissues and
that of the venous supply to remove the generated wastes.
Perhaps the most obvious differences therefore will be in
the blood gases, oxygen and carbon dioxide, and as a consequence on bicarbonate ion concentration and pH. Food
materials such as glucose and wastes such as urea will also
differ in concentration in the two types of blood. In general,
differences in enzyme activity are less easy to predict but in
any case there is rarely any necessity to study the activities
in arterial blood.

5.1.3. The Processing of Blood Samples

Two principal types of processed blood specimen are used in laboratory analyses: (*a*) plasma which is blood with an anticoagulant
added to prevent clotting and (*b*) serum which is the fluid left after
a clot has been formed naturally.

(*a*) Plasma

It is unfortunately the case that some commonly used anticoagulants
have a number of effects on the specimen. Since many of them act
by removal of the calcium ions required for clotting, the consequential ion changes can result in water movements between cells and
extracellular fluids giving a dilution or concentration effect. Sodium
and potassium EDTA seem not to cause such changes but it is generally considered that if plasma is to be used then heparin is the
best of the available anticoagulants.

Some anticoagulants also inhibit enzymes directly; fluoride is perhaps the most well known and since it inhibits glycolytic enzymes present in red blood cells, it is actually useful in preventing the metabolic removal of glucose in blood samples. Removal of required cofactors, especially Mg^{2+} ions by chelation-type anticoagulants is another means of enzyme inhibition, the enzyme creatine kinase being particularly susceptible to such effects. A final problem is that occasionally anticoagulants seem to interact with various components of the enzyme assay system itself, eg with oxalate ions and the assay for acid phosphatase being such a case. This can make the comparison of results obtained using different assay methods quite difficult.

(b) Serum

A logical development of this problem is to allow the blood to clot and use the serum thus obtained. Notwithstanding the fact that the large size of the enzyme molecules means a number of unfortunate effects can occur as a corollary of the clotting process, this method is indeed commonly used.

One effect of clotting is that some enzymes seem to show a reduction in activity in serum even after allowing for the differences in volume compared with whole blood, probably as a result of physical entrapment within the developing clot or spontaneous attachment to the clotting proteins. Other enzymes show a dramatic rise in activity due to the release of intracellular enzyme molecules from those platelets and blood cells which disrupt as the clot forms. Acid phosphatase shows this effect and, despite the fact that the enzyme molecules released are sometimes sufficiently different to be distinguishable by differential inhibition (eg by tartrate in this case) or by careful choice of substrate, it is wise to remove the serum from the clot as soon as possible. In some instances these effects can be very subtle and in the case of CK, haemolysis during clotting affects the apparent enzyme activity by virtue of the release of adenylate kinase (AK), an enzyme which will alter the concentration of ATP by the reaction:

$$2\,ADP \rightleftharpoons ATP + AMP$$

∏   Refer to the assay procedure for CK (Section 3) and in par-
      ticular to Fig. 3.2j. How might the presence of AK adversely
      affect the assay?

As part of the assay system the two substrates ADP and creatine phosphate are added in fairly high concentration. The
presence of AK will reduce the level of ADP by the above
reaction, but if the ADP concentration is sufficiently high
this may not produce a noticeable decrease in CK reaction
rate. The most significant effect of AK will be the production
of ATP since this will be in addition to the relatively small
amount produced by the activity of CK. Both the $NADP^+$
coupling system shown in Fig. 3.2j and the newer luminometric approach will respond to this ATP and thus give falsely
high assay values (Fig. 5.1a).

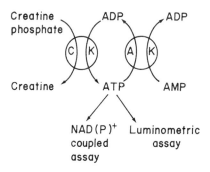

Fig. 5.1a.   *Scheme to show the potential interference of AK
with the CK enzyme system*

A sample treatment that in general is desirable in chemical analysis
is the removal of proteins using various precipitating and denatur-
ing reagents. Pause for a moment and ask yourself why this is not
possible in enzyme assays.

The answer is that enzymes are themselves proteins, as succinctly
expressed in Schardinger's famous statement 'All enzymes are proteins but all proteins are not enzymes'. Thus any technique for the
removal of the general mass of protein is very likely to affect the
enzyme of interest.

∏   (*i*)   Think in general terms about the assay methods discussed in Section 3 and see if you can predict what problems this restriction on deproteinisation might produce.

      (*ii*)   Accepting that deproteinisation is not feasible and that we must therefore assay enzymes in the presence of many other enzymes and non-enzymic proteins, what is it about enzymes that enables us to do this satisfactorily in most cases?

      (*i*)   Proteins are very sensitive to inhibition and denaturation, and many chemical reactions of potential value in the measurement of the product of a reaction contain one or more components, or have an aspect of the reaction environment (pH extremes, high temperature etc), which will result in inhibition or denaturation. As explained earlier, this means that continuous measurements produced by inclusion of the derivatizing reagent in the assay system are not possible. Either discontinuous approaches will be needed, or some means such as dialysis employed to separate the enzyme from the low mass substrates or products which are to be measured. Alternatively steps must be taken to remove insolubilised or denatured protein before the final measurement can be obtained.

      (*ii*)   In order to measure a given enzyme at all in the presence of many others we rely upon the high substrate specificity that most possess, but of course it does become difficult to measure the relatively unspecific kinds of enzyme in this situation.

5.1.4. Stability of Blood Sample Components and Sample Treatment

(*a*) *Enzyme Stability*

The stability of enzymes in biological samples varies substantially and some prior knowledge of the enzyme to be analysed is necessary in order to ensure optimal storage conditions. Whereas cholinesterase (CHE) and amylase (AM) are relatively stable for at least a week even at room temperature, sorbitol dehydrogenase (SDH) loses about 20% of its activity per 24 hours even when refrigerated. Enzymes with isoenzymic forms can show variation in stability between the different isoenzymes and while this can act as an unfortunate error-producing variable, it also provides a means of distinguishing them as will be discussed in Part 9. Perhaps most well-known and important of such differences occurs with LDH where the so-called type 5 (predominantly of liver origin) is relatively unstable at temperatures below 4 °C and above 45 °C, whereas the type 1 (of heart origin) is comparatively stable up to 65 °C.

Stability can be improved in a number of ways, some of which are fairly obvious such as storage at low temperatures. However even with such a simple procedure subtle problems can arise since a number of enzymes which contain several protein chains are in fact *less* stable when frozen than at 4 °C. A range of other stabilising treatments is available including the use of protective agents such as thiol compounds to maintain the thiol groups of the enzymes in a reduced state. Acid phosphatase is interesting in being significantly more stable when the pH is lowered to 5.5–6.0, surviving for about a day at 25 °C, whereas at pH values above 8 its activity disappears within an hour or so. Many other enzymes have unusual and unpredictable requirements for maximum stability.

(*b*) *Enzyme Inhibition*

Inhibition of enzyme activity by extraneous materials must be reduced if at all possible because of the serious effects of these materials and the considerable and unknown variability of the effects in question. Contamination of apparatus, containers and solutions by heavy metal, powerful detergents and the like has to be rigorously reduced.

(c) An Example of the Problems Encountered

It is hoped that the previous text has shown that enzymes vary substantially in their requirements and in the problems they create, and that no universal recipe, regime or advice can be given for optimal sampling, transport, storage or assay. Creatine kinase will act as a useful summary or example of the problems encountered although fortunately it is a rather extreme case.

CK is:

— subject to diurnal changes – but only in females! Commonly the activities in samples taken in the late afternoon are over 60% higher than those in early morning samples;

— present in activities as little as 30% of normal in the later stages of pregnancy;

— markedly influenced by surgery and exercise, the dramatic rises obtained taking as long as 5–10 days to return to normal for that individual;

— inhibited by EDTA, citrate and fluoride which restricts assays to serum or heparinised plasma;

— relatively unstable; taking 4 h at 20 °C, 8–12 h at 4 °C and 2–3 days at less than 0 °C for its activity to decline by 50%. Fortunately the rate of degradation can be slowed by the addition of reducing compounds containing thiol groups;

— subject to inhibition from an unknown serum factor which seems to be present in variable amounts;

— assayed by a method which is subject to interference by materials that can be released from haemolysed red cells.

SAQ 5.1a

(*i*)   State *four* examples or situations in which biological specimens including routine blood specimens, might only be available in small quantities.

(*ii*) Very briefly describe *two* ways in which we can increase our ability to measure the enzymes in these small samples.

SAQ 5.1b

Adenylate kinase, released from cells which become damaged during or following the taking of blood samples, interferes with creatine kinase assay, (Fig. 5.1a). How do you think the release of another type of enzyme, the virtually omnipresent ATPases, would affect this assay? They carry out the following reaction:

$$ATP + H_2O \rightarrow ADP + PO_4^{3-} + H^+$$

Try and answer this question from memory before referring to Fig. 3.2j, in which the CK assay system is illustrated.

SAQ 5.1b

SAQ 5.1c   Briefly describe *two* of the possible effects on enzyme assays of the use of anticoagulants.

5.2.   OTHER BODY FLUIDS

In principle enzyme activities could be measured in a specimen from any part of the body but in practice samples other than blood are used relatively infrequently.

5.2.1.   Urine

Measurements of enzyme activity in urine present a number of analytical problems including:

(*a*) the presence of activators and inhibitors in variable concentration;

(*b*) the intermittent nature of urine voiding and the variation in dietary liquid intake, which lead to considerable variation in urine output and a consequential dilution or concentration effect;

(*c*) the tendency for some enzymes to be have a diurnal variation in production;

(*d*) the fact that drugs can affect enzyme levels and activities, usually resulting in a reduction in measured activity, although the common antibiotic gentamycin can result in a substantial increase in activity of some enzymes.

The problems in (*b*) and (*c*) can be reduced in importance by the averaging effect of collecting complete 24 hour urine outputs. However some enzymes tend to degrade during this long period, especially if the samples are stored under poor conditions, which is often the case if the collection is carried out by the patient at home. The use of preservatives in this situation is fraught with problems due to the sensitive nature of enzymes.

Of the half dozen or so enzymes measured with any frequency in urine the majority are used to investigate non-kidney conditions. The rise in output of amylase as a result of pancreatic inflammation is a typical example. Some however may highlight important conditions in the kidney itself and particularly interesting is the use of N-acetyl-β-D-glucosaminidase; an increase of which indicates tubule damage and may give 2–3 days warning of impending kidney transplant rejection.

5.2.2. Faeces

Faecal enzyme levels in adults are very low but much more significant in children perhaps due to the more rapid throughput of faecal matter. While contamination with bacterial enzymes is something of a problem the low level of pancreatic enzymes in the faeces is an important diagnostic feature in cystic fibrosis in very young babies.

5.2.3. Cerebrospinal Fluid (CSF)

Some correlations exist between enzyme levels and infection, infarction and haemorrhage but they add little to the information derived from samples obtained by other easier and safer means. A reliable correlation with cerebral tumours would be extremely valuable but has not yet been found. As a consequence of these deficiencies enzymes are rarely measured in CSF.

5.2.4. Biopsy Specimens

An increasingly important area of work concerns the measurement of enzymes in tissue samples taken as biopsies. Again inherited diseases are commonly the stimulus for this, although a wide range of investigations on fetal well-being are undertaken using such biopsy samples. In many cases the biopsy will produce only a relatively small piece of tissue and even after an increase in its cell number by means of tissue culture techniques the sample is often still sufficiently small to raise major problems in analysis. Of course this is one of the incentives for carrying out the investigations by means of enzyme assay in the first place since the catalytic nature of enzymes should make a measurable result more likely by virtue of their recycling action.

Of the wide range of tissue specimens that could be taken those from the fetus or its immediate environment are perhaps the most interesting. Enzyme activities can be measured in the fluid phase or the cells of the amniotic fluid, or in the blood or cells from the fetus itself.

It is usual to increase the number of cells obtained in the original sample by cell culture techniques but this can generate problems (Young *et al*, 1975):

— the end culture may be descended from relatively few, perhaps atypical, cells;

— the act of cell culture alters enzyme types and levels. Low half-life enzymes are usually found at higher activities in cultured cells due to their rapid growth and division compared with that of normal body cells. Lysosomal enzymes are particularly variable although if expressed as a ratio with another such as β-galactosidase, less variation is seen;

— enzyme patterns can change with growth conditions, particularly pH, carbon source and the concentrations of salts, vitamins and the level of CO_2;

— the enzyme types and activities vary with the cell type being cultured and some cultures from amniotic fluid are mixtures of epithelial and fibroblast cells. Unfortunately these differ quite a lot and attempts are usually made to manipulate the conditions to obtain relatively pure fibroblast cultures.

As might be expected with an area as new as this a number of fundamental problems are encountered particularly if investigations of activities in the cells are envisaged (Latt and Darlington 1982). A fairly fundamental but important point is the current poor state of control data, ie the expected activities, acceptable ranges etc, in the normal fetus.

Notwithstanding the increase in cell number obtained using these techniques, very great sensitivity and scale-down is required of the enzyme assay procedures. Fluorescent and radioactive substrates together with special techniques that cannot be discussed here are frequently employed. See Wooton and Freeman (1982) for a detailed discussion of microanalytical procedures.

Many other examples of the use of the biopsy technique are available, and to give just one illustration, small samples of duodenum or jejunum can be removed from the gut using a wide bore needle and then assayed for the enzyme lactase. This can be done by preparing a tissue extract and then using the types of assay technique described earlier (Part 3), or more interestingly by a visual histochemical technique. In this latter approach sections are cut and then treated with assay reagents which allow the lactase to generate a coloured derivative. The absence of the (usually green) colour in the cells of the gut wall when examined microscopically is suggestive of the presence of the inherited disease 'lactase deficiency' in the patient.

Sophisticated and sometimes complex techniques for sample collection are available; these attempt as near an instantaneous cessation of metabolic and spontaneous events in cells as is practicable in order to isolate as 'natural' a sample as possible. These usually involve deep frozen clamps and cutters to freeze the selected tissues very rapidly. Such techniques are more appropriate to research investigations and in any case are more usually required to arrest changes in very labile intermediates such as nicotinamide coenzymes, ATP and its derivatives, intermediates of metabolic pathways and the like, rather than in studies on enzymes. They are in general too time consuming, skill requiring and in fact generally unnecessary in the routine clinical chemistry laboratory.

Summary

This part briefly surveys the main types of biological specimen taken from patients for analysis but concentrates on the factors that affect the specimen during or subsequent to its removal. In particular the influence of some concurrent physiological events, changes during the immediate processing of the specimen and some general aspects of the stability of sample enzymes are considered.

Objectives

You should now be able to:

- briefly state the importance of the sampling process to the value of enzyme analysis as an aid to clinical diagnosis;

- list at least four different human sample types that are used in clinical enzyme assays;

- discuss the chemical differences between arterial, venous and capillary blood;

- compare the merits and problems of using serum and plasma for enzyme assay;

- give examples of factors affecting enzyme stability and activity in blood samples;

- list at least three problems associated with the analysis of enzymes in urine;

- give one example of the usefulness of fetal, faecal and cerebrospinal samples in clinical investigations;

- discuss the problems associated with enzyme analyses of cell cultures derived from human biopsy specimens.

6. Practical Aspects of the Measurement of Enzyme Reaction Rates

Overview

In this Part some important practical aspects of the measurement of enzyme reaction rates will be discussed. In particular we will be concerned with the number of time points used to determine the quantity of reaction product generated, and the methods used by automatic analysers to identify the linear parts of the reaction profile from which the rate calculations need to be made.

The previous parts of this Unit should have convinced you that enzyme reactions are best measured by following the appearance of reaction product during the initial stages of the reaction. In this part we will examine how this is achieved in practice, and discuss some of the problems and merits of the various procedures commonly used.

6.1. INTRODUCTION

Fig. 6.1a shows the profile of a typical enzyme reaction and it can be divided into two regions. Region 2 appears when the reaction ceases due to substrate depletion, enzyme inactivation or the other events discussed in Part 1.

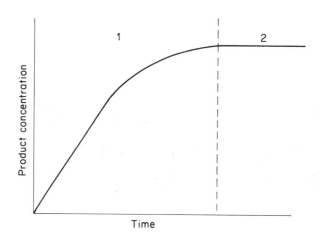

Fig. 6.1a.   *A typical enzyme reaction profile*

Since enzyme levels have no effect on the numerical values obtained in that region, methods (equilibrium, steady-state or end-point methods) which depend upon measurements there, are only really suitable for the assay of chemical concentration and not for the assay of enzyme activities. It is actually possible to use that region to assay enzyme activities by measuring the time required to reach the plateau, since this time will decrease with increasing enzyme activity. However, since determination of this time point is difficult, measurement error will be high and furthermore each assay is likely to take a reasonable length of time. In fact this time could be quite long for low activity samples, and the method is difficult to carry out with large numbers of samples if their enzyme activities and hence reaction times are very variable. Such procedures are not popular in laboratories since this variation in reaction time makes work organisation quite difficult and furthermore they are not really compatible with the construction and mode of operation of most automatic analysers in which samples progress through the assay modules at fixed speeds.

The term 'end-point method' is best avoided for these approaches since some workers have used this term for methods in which the assay has been *ended* by the addition of a material, change of conditions etc, when in fact the reaction may still be in region 1 of the graph. To avoid confusion, 'equilibrium' or 'steady-state methods' are better terms.

Data acquired within region 1 of the curve are more commonly used in enzyme rate assays. Methods based upon this approach are frequently called 'kinetic' methods, which is a term that arose to imply that the reaction is proceeding rapidly, but it is strictly incorrect since chemical changes are still occurring even in region 2 of the reaction profile.

∏ In advance of our discussion can you imagine what are the various advantages that assays using region 1 of the profile have over the equilibrium methods. These advantages reflect the importance to hospital laboratories of carefully organising their daily work load, but they also have significance regarding the quality of the results obtained.

(*a*) One immediate advantage is the speed of the assay since measurements can be taken over a period as short as one minute in some cases. No matter what length of time is actually required to obtain a reliable measurement of the rate during the linear part of the profile, it is bound to be shorter than if one has to wait until the equilibrium is reached, and more importantly, is confirmed.

(*b*) Measurement over the early part of the profile gives maximum discrimination between alternative reactions as shown in Fig. 6.1b. Curve 1 represents the profile shown by the reaction of interest whereas curve 2 might represent the same enzyme in another sample.

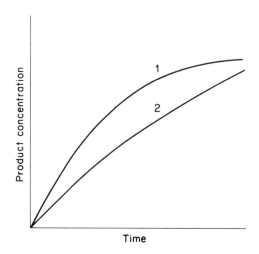

Fig. 6.1b.   *Reaction profiles of enzyme (1) and interfering
(2) reactions*

It is apparent that the two lines show the greatest difference
in rate (ie slope) in the early part of the graph. Curve 2
might also represent the changes due to interfering reactions
produced for example by other enzymes in the sample, or
chemical effects on the assay components.

(*c*) Since the object of taking measurements in this early period is to measure changes in concentration, the actual
numerical values are not important, a change from 0.1
to 0.2 units being as meaningful as a change from 0.15
to 0.25 units. The significance of this is that the starting value for the reaction is of little importance, hence
measurements of the activity of turbid or coloured specimens is comparatively straightforward.

(*d*) The initial reaction rate is more reproducible and reliable than rates obtained later in the profile because
enzyme denaturation and inhibition effects should be
minimal, and substrate and cofactor depletion should
not have occurred even with highly active samples.

SAQ 6.1a	Fig. 6.1c shows typical reaction profiles for:

(1) a turbid sample;
(2) a normal, clear sample.

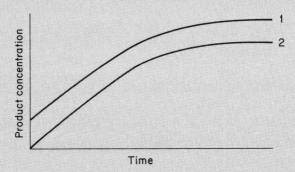

Fig. 6.1c. *Reaction profiles for* (1) *a turbid sample, and* (2) *a normal, clear sample*

(*i*) Use this figure to describe one of the advantages of assay systems which depend upon the measurement of reaction rates during the early part of the reaction.

(*ii*) List the other advantages.

6.2. THE NUMBER OF MEASUREMENTS TAKEN IN THE ASSAY

With regard to the design of experimental methods for the measurement of initial reaction rates, the major variable is in the number of data points collected for the purpose of calculation.

6.2.1. One-point Assays

The simplest approach is to take a single (usually absorbance) measurement at a pre-determined time, t_x (Fig. 6.2a); the value of this measurement being converted to product concentration by a calibration curve or correction factor.

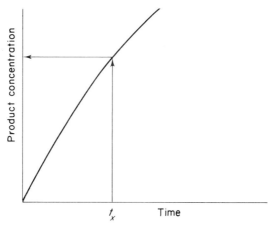

Fig. 6.2a. *One point assays of enzyme reactions*

∏ If this method is to provide an accurate determination of reaction rate, what are the assumptions that must be made about:

(*a*) the initial absorbance (or other measurement) value, and

(*b*) the shape of the response curve?

Try sketching a few alternative graphs and look at the effects on the y axis value obtained.

(*a*) It must be assumed that the starting value is either zero or at least is of a constant value. A sample blank will be needed.

(*b*) Similarly a linear response over the period up to t_x is required, which implies that the time before the measurements are taken should be as short as practicable. This may mean that only a small change in measurement signal will have occurred with a consequently larger potential for error.

In practice there are perhaps two different technical approaches used to obtain this measurement. In manual methods, and in the so-called discrete automatic analysers, which generally mimic quite closely the manual procedures, the reaction is usually stopped at t_x by the addition of inhibitors, colour or derivative developers or simply by dilution with excess reagent. Such approaches are called interrupted methods. The second approach is also found in automatic analysers and involves the flow of reagents and sample through tubes (the continuous flow systems). In this case the sample flows past or through a sensing device without stopping. The time delay for the reaction can be altered by adjusting the flow rate or the position of the sensor in the flow line. The construction, use and merits of these analysers are beyond the scope of this text.

6.2.2. Two-point Assays

While one-point assay methods are certainly used, a much more common approach to reaction rate measurement involves two measurement points, with the calculation being based upon the change occurring between them (Fig. 6.2b).

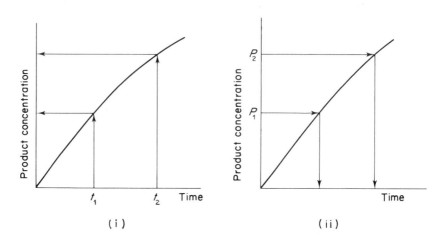

Fig. 6.2b. *Two-point assays of enzyme reactions* (i) *fixed time points;* (ii) *variable time points*

In general two pre-selected time points t_1 and t_2 are used and the signal (usually an absorbance measurement) at each is recorded (Fig. 6.2b(i)). In some situations two signal values P_1 and P_2 are chosen and the time required to change between them is measured (Fig. 6.2b(ii)). Such 'variable time' methods are relatively uncommon as they are more difficult to automate.

Two-point fixed-time methods are very common in clinical laboratories both in manual assay procedures and in automatic reaction rate analysers of the simpler kind. Their popularity is probably due to their simplicity, especially of the data processing, but they also suffer from some significant problems.

∏   Consider Fig. 6.2b(i) for a moment and attempt to identify some of the main problems involved.

The most important problem is undoubtedly that of ensuring that a constant reaction rate is obtained for a reasonable length of time; ie the reaction has zero-order kinetics (so that the rate is independent of substrate concentration) between the chosen time points. It is therefore necessary to know:

(a) the presence and extent of any lag period in the reaction profile;

(b) the presence and extent of any initial high reaction rate due for example to the metabolism of endogenous substrate added with the sample, or contaminants in the experimental substrate acting as substrate analogues;

(c) the effect of sample activity, drug treatment etc on both (a) and (b) is also an important factor;

(d) the effect on the linearity of any variation of sample activity over the expected range, in both normal and pathological specimens, eg highly active samples will probably show an early substrate depletion.

An important point is that dilution of the sample in order to reduce the enzyme activity and extend the linearity can generate further problems. For example some enzymes are more stable at higher concentrations than when diluted, or more stable in complex mixtures (such as serum) than in simpler ones (such as diluted serum, or simple buffer solutions). Furthermore dilution will reduce the level of enzyme activators or inhibitors present in normal serum and result in a relative change in activity. In other words the enzyme activity in a diluted specimen may not be in direct proportion to that in the concentrated original sample notwithstanding any change in actual activity of the enzyme molecules as a result of dilution. Additionally the dilution step will increase assay time and potential error.

Therefore a careful choice of assay times is necessary and, ideally they should be varied for different specimens. However advance knowledge of sample activity is unlikely and hence repeat assays may be necessary, although this would be very unpopular due to the time and cost involved. Furthermore many autoanalysers are not amenable to easy and regular changes in timing and the associated calculations.

In practice therefore the first time point is selected as one that is likely to be beyond the possible initial lag or high activity regions for most samples to be assayed. The second one is chosen to allow sufficient time for a reliable measureable change to occur while being well within the expected linear portion for most samples.

The actual technical approach to two-point fixed-time assays differs between discrete or continuous instrument designs. In neither case can a reaction stopping treatment be used for the first time point, so in general unless direct measurement of the reaction product is possible, discrete analysers (eg the LKB 8600 Reaction Rate Analyser) remove samples at selected time intervals to separate modules containing stopping or derivatizing reagents. On the other hand, in continuous systems (eg the Technicon SMAC) the reacting mixture flows past or through two sensing devices spaced a suitable distance (and hence time) apart. The development of this apparently simple principle was only possible when devices were designed in which the pronounced sensor signal produced by the passage of the bubbles which drive the solutions along the tubing, could be eliminated electronically. This flow-past approach overcomes the problem of timing the reaction from its initiation to its termination since it is the time delay between the two selected points that is used in the rate calculations. For continuous systems this is quite important since the precise time of completion of mixing of sample with initiating or terminating reagents is difficult to determine and engineer reproducibly.

∏   Why do you think careful control of flow rate is important in these continuous flow systems?

In a system where measurements are taken as the samples flow past two spaced sensors, the reaction is continuing during the time it takes to flow between the sensors. Since a change in flow rate will change the time taken for this travel, and it is unlikely that an allowance will be made in the calculation of reaction rate for such variation, errors will be introduced if the flow rate varies significantly. Attempts must be made to stabilise flow rates within acceptable limits.

Having been very negative about the two point assays it is worth pointing out that the method does have the advantage of relative simplicity, and reducing or even eliminating the need for blank specimens, since it is the change occurring between the two time points that is significant. As pointed out earlier the consequence of this is that the starting value is not usually important and thus blank subtraction need not be carried out. It is the case that this general two-point approach is the most common among the manual methods and the simpler automatic analyser systems.

SAQ 6.2a

Take each of the following brief descriptions of enzyme assay methods and identify the type of measurement being undertaken (one-point, two-point, fixed-time, variable-time etc).

(*i*) *Alkaline phosphatase*
Sample is incubated with phenolphthalein mono-phosphate. After 20 minutes 0.1 mol dm^3 pH 11.2 buffer solution is added to inactivate the enzyme and increase the colour intensity. Measurement is made at 550 nm against reagents processed without sample.

(*ii*) *Caeruloplasmin I*
Sample is incubated with o-dianisidine and portions are removed at 5 and 15 minute intervals and added to sulphuric acid. Absorbance measurements are made at 540 nm against water.

(*iii*) *Caeruloplasmin II*
Sample is incubated with a substrate in a spectrophotometer cuvette and changes in absorbance at 460 nm are recorded on a chart. The reaction rate is calculated from the slope at zero time. $\longrightarrow$

(*iv*) *Lactate dehydrogenase*
Sample is incubated with lactate and NAD^+ for 7 minutes. An acidic solution of cuprous (copper(I)) ions and neocuproine is added to stop the reaction and develop the colour. The absorbance is measured at 455 nm against a sample of serum.

(*v*) *Aspartate aminotransferase*
Sample is incubated with a complex substrate mixture in a spectrophotometer cuvette. A short time is allowed for the lag period to pass and the reaction rate analyser collects 5 absorbance measurements at 0.1 s intervals. This is repeated after 6 s and the difference between the two sets used for the rate calculation.

(*vi*) *Alkaline phosphatase*
Sample is incubated with disodium-p-nitro-phenyl-phosphate and the enzymic release of p-nitrophenol is measured. The analyser is set to record the time when transmittance passes 90% and 88.2% values, and the time difference is used to calculate the reaction rate.

SAQ 6.2a

SAQ 6.2b   List three potential problems with choosing time points t_1 and t_2 for the initial and final assay points in a typical two-point enzyme assay procedure (Fig. 6.2b)

SAQ 6.2c   Graph (i) of Fig. 6.2c represents an enzyme reaction profile with a lag period and an equilibrium plateau. Graph (ii) represents a reaction without a lag period but which proceeds at the same rate.

Use each graph to determine the reaction rate (1) after 4 minutes and (2) between 2 and 4 minutes. Comment on the results obtained. What are the names given to the measurement approaches in (1) and (2)?

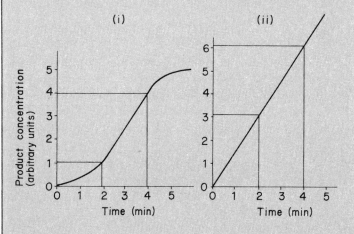

Fig. 6.2c.   *Two-point assays of enzyme reactions*

SAQ 6.2d	A highly active sample shows an apparent activity of 100 U cm^{-3}, on dilution to 50% the activity becomes 40 U cm^{-3}. Explain why the latter value might be obtained instead of 50 U cm^{-3}.

6.2.3. Multi-point Assays

A logical development of the two-point approach is to try and increase the number of time points used. It is difficult to produce a large number of timed measurements in manual assays and in the continuous flow systems. Even the Technicon SMAC, which is probably the most complex of the latter devices, has a maximum of only three time points.

One development that did attempt to increase the number of time points in a continuous system involves the pumping of a sample/reagent mixture into a heated coil. When it becomes full a valve reverses the flow to evacuate the coil and consequently the first segment of mixture to enter the coil is the last to leave, and has therefore been incubated for the longest time. A single photometric detector at the coil inlet is thus capable of giving many measurements over a wide time span. The precision of timing however proved inadequate for reliable results.

Discrete systems can be easily constructed to take multiple measurements, and with powerful microprocessors being so readily available, the handling of the data acquired is comparatively easy nowadays. The term 'kinetic assays' for those involving multiple measurements is frequently seen, but is inappropriate for the reasons mentioned earlier (p. 103) and the term 'continuous assay', while even this is not strictly correct, is perhaps more suitable.

Commercially available automated enzyme reaction rate analysers vary considerably in the number of data points they collect and the time interval between them. The Ken-o-Mat system produced by Coulter Electronics Ltd only acquires absorbance readings at 64 second intervals for 5.5 minutes, and the Demand system of Cooper Biomedical Inc uses 36 seconds intervals. The Hitachi 705 collects measurements every 20 seconds for 10 minutes, and the Kodak Ektachem 700 every 6 seconds for 5 minutes. Of the 'manual mimic' type of discrete system the American Monitor Corporation Parallel is particularly good in using 8 data points acquired over a 10.5 second time period, with each datum point being an average of 4 separate readings.

A very important type of discrete analyser is the centrifugal analyser, in which the enzyme reactions are carried out along the radii of a spinning disc (rotor) with the spectrophotometric measurement being taken while the sample and reagents react in a special chamber at the end of each radius (Fig. 6.2d)

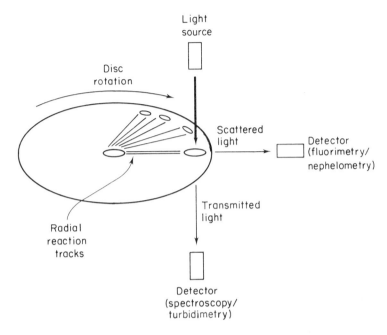

Fig. 6.2d. *Schematic representation of centrifugal analyses*

Centrifugal analysers have a number of distinct advantages some of which are associated with the spectroscopic assay of the developing reaction.

(*a*) The fact that the various reactions run along adjacent radii of the rotor means that they will operate under identical and parallel time conditions and hence accurate timing of reaction initiation and assay is of minor importance. This is particularly useful for rapid reactions.

(*b*) While the instruments have the cost and low error benefits of single beam optics, a double beam measurement is effectively created by the regular monitoring of a reference solution which gives 100% transmittance, with a zero transmittance value being obtained from the solid non-transmitting parts of the rotor between each reaction radius.

(c) Perhaps the most significant spectroscopic advantage of these systems, however, is the shear number of data points that they are able to collect, as an example taken from Burtis and Mrochek (1980) will show.

When a 13 cm diameter rotor with 15 analytical and 1 reference track spins at 4800 rpm, each revolution takes 12.5 ms and each of the 16 cuvettes (0.32 cm diameter) occupies the sensing light path for 98 μs. This time span is ample for measurements to be taken and results in the acquisition of 1280 data points per second; a great improvement on 1, 2 and 3 point assays! Unfortunately the microprocessors involved with these systems cannot usually cope with this data rate and to avoid memory bank overload usually tend to be selective and record fewer data points.

Some systems carry out data reduction by signal averaging, whereas others select the highest value obtained in that cuvette pass. The American Instrument Corporation Rotochem is constructed so that each time point of 1.3 seconds involves 8 revolutions of the rotor, and the 8 absorbance values are summed to provide a datum point. In general the centrifugal analysers currently available produce between 10 and 100 usable data points per second for the rate calculations.

∏   Going back to the example given earlier, consider a 13 cm diameter rotor containing 16 cuvettes of 0.32 cm diameter. If rotor speeds are (i) 1800 rpm and (ii) 500 rpm, calculate:

(a) the total number of data points acquired per second.

(b) the measuring time per cuvette per revolution.

Let us first of all work through the example given in the main text.

At 4800 rpm the rotor will take 60/4800 = 12.5 ms for each revolution and will perform 4800/60 = 80 revolutions per second. Assuming only 1 datum is collected per cuvette per revolution, then the number of data collected per second is

$80 \times 16 = 1280$. With a 13 cm diameter rotor the circumference will be $2 \times \pi \times 6.5 = 40.84$ cm and each cuvette will occupy $0.32/40.84 = 0.784\%$ of this. Thus if 12.5 ms are required for each revolution, each cuvette must be analysed in only $12.5 \times 0.783 = 97.9$ μs!

The calculations for the other rotor speeds are as follows:

Revolutions (m^{-1})	Revolutions (s^{-1})	Rate of collection of data (s^{-1})	Time per revolution (ms)	Measuring time per cuvette (μs)
4800	4800/60 = 80	80 × 16 = 1280	60/4800 = 12.5	12.5 × 0.784 = 97.9
1800	1800/60 = 30	30 × 16 = 480	60/1800 = 33.5	33.3 × 0.784 = 261
500	500/60 = 8.3	8.3 × 16 = 133	60/500 = 120	120 × 0.784 = 941

The ability of these systems to carry out very early measurements, in many cases less than 4 seconds after rotor acceleration initiates the reaction by causing the sample and reagents to mix, has a significant number of advantages.

(*a*) In some centrifugal analysers it is assumed that for normal reaction rates the first measurement taken is of unreacted sample and can therefore be considered as a sample blank. Although a reagent blank may still be needed, this so-called auto-blanking approach saves considerable time and effort.

(*b*) As pointed out previously early measurements also decrease the effect of interfering reactions since most such reactions are relatively slow to develop and their influence is consequently minimised.

(*c*) An important organisational point is that the rate of sample
processing can be very high in these systems, so for example
the Union Carbide Corporation Centrifichem can process up to
180 samples per hour, whereas the non-centrifugal LKB 8600
Reaction Rate Analyser can only manage about 50 samples per
hour.

The reliability of such short term measurements is of great interest,
bearing in mind the relatively small absorbance changes that will
have occurred and have to be distinguished. A point worth noting in
this context is that the quality of photometric devices in reaction rate
analysers is nowadays extremely high. Perkin–Elmer devices have
for instance been shown to have a standard deviation of repetitive
measurements of less than 10^{-5} absorbance units and the Du Pont
ACA system is said to have a spectrophotometric sensitivity of 0.28
milli-absorbance units.

SAQ 6.2e	List the advantages of centrifugal analysers for the assay of enzyme reaction rates.

6.3. CURVE DRAWING AND IDENTIFICATION OF THE LINEAR REACTION RATE REGION

Determination of an accurate reaction profile and identification of the linear part within it, is of considerable relevance to the reliability and meaning of the results. One immediate advantage of obtaining many data points is that if manual drawing of reaction profiles is undertaken, production of an accurate curve becomes easier and the linear part ought to be more apparent. Reliable curve fitting coupled with the availability of many data points is particularly important for very fast reactions and has for example resulted in a fall in the intra-laboratory coefficient of variation for the so-called EMIT assays, from about 15% for manual methods to less than 2% when carried out by centrifugal analysis.

However the ability of microprocessors to handle prodigious numbers of mathematical calculations at high speeds has led to the development of more sophisticated techniques. The fitting of the best (curved) line to the data obtained becomes relatively easy provided the expected curve is known and the appropriate statistical algorithm available. The Coulter Ken-o-Mat uses a least squares approach to determine the line of best fit through the calibration points and notifies the operator if it does not agree, to within 2%, with the slope of the line through the highest point.

A number of approaches are used in the various microprocessors to determine linearity, of which a relatively simple and common one is to determine the slopes (δ) over the periods between the time points, compare them and use the portion of the curve where adjacent slopes are equal (Fig. 6.3a).

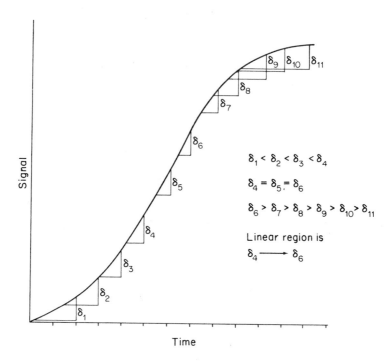

Fig. 6.3a. *Determination of linearity by the delta (δ) method*

Rate calculation is then comparatively straightforward since, as a prelude to calculating the slopes, the absorbance changes over each region will have been obtained and are available for the calculation. In general if a small number of data points are present in the linear region (which would make the calculations less reliable), then the assay is abandoned by the microprocessor. For the Coulter Ken-o-Mat five such points must be present.

Another slightly different approach is to calculate the differences between the adjacent regions, average these, and use these values for the calculation. In other devices the area under the curve is calculated over each time span rather than the slope and the difference used as a basis for the rate calculation, a so-called integral approach (Fig. 6.3b)

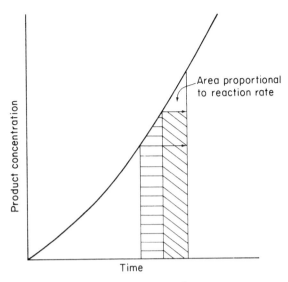

Fig. 6.3b. *The integral approach to rate measurement*

Instead of determining slopes at the various time points used for the measurement, slopes can be drawn at many points along the reaction profile once a smoothed curve has been fitted. A comparison of the large number of slopes obtained is made as described earlier.

∏ Absorbance data from an enzyme reaction are shown below, use a slope technique to identify the beginning of the linear region.

Time of measurement (t) (s)	Absorbance (A)
1	0.100
2	0.100
3	0.100
4	0.110
5	0.140
6	0.185
7	0.250
8	0.420
9	0.585
10	0.750

A simple slope approach would involve the following calculations which show that the linear region begins about seven seconds after the start of the reaction. You could draw the curve yourself to emphasise this point.

Time lapse (Δt)	Absorbance change (ΔA)	Slope ($\Delta A / \Delta t$)
1–2	0	0
2–3	0	0
3–4	0.010	0.010
4–5	0.030	0.030
5–6	0.045	0.045
6–7	0.075	0.075
7–8	0.165	0.165
8–9	0.165	0.165
9–10	0.165	0.165

In many instruments a more complex approach to the identification of linearity is taken and a short study of the method used in the Cobas BIO makes an interesting example.

This device carries out its analysis in four stages.

— The absorbance profile is checked to find the linear regions, and if less than 4 data points are present in this region then the assay is abandoned and the operator notified.

— The optimum line is fitted to the data using a linear regression program.

— A calculation is made of change in absorbance per minute.

— Enzyme activity is calculated using an appropriate calibration factor.

The detection of the linear region involves the identification of its end-point and then of the starting point of the region using a technique involving comparison of triangulation angles.

Let us take an example in which the absorbance is rising during the reaction, and absorbance data are being collected at selected time intervals beginning with a t_0 value following a period of pre-incubation without sample or an essential reagent.

The micro-processor selects a time point (the auxiliary point end, APE) which is five seconds in advance of t_0 and 'draws' lines to each absorbance value obtained, before calculating the angle made to the horizontal line formed at A_0 (Fig. 6.3c). Beginning with the final absorbance point obtained it compares the angles generated with each previous one. When an obtained angle becomes less than the one previously calculated, then this latter point is regarded as the end point of the linear region. Thus in the profile illustrated:

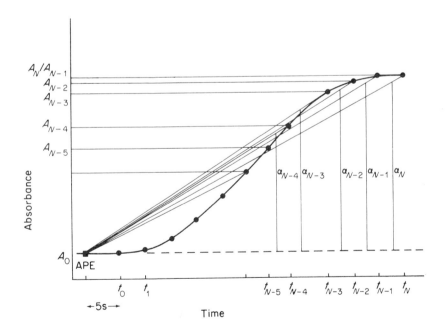

Fig. 6.3c. *Determination of the end of linearity*

$$\alpha_N < \alpha_{N-1} < \alpha_{N-2} < \alpha_{N-3} > \alpha_{N-4}$$

and absorbance N–4 must be regarded as the end of the linear range of the curve.

∏   To determine the start of the linear region the reverse procedure is used. Perhaps you would like to attempt to describe this procedure using the following stages.

— The selection of the point (called the auxiliary point beginning, APB) which will be used for the angle determination.

— The line drawing and the sequence of angle comparisons.

— The differences between adjacent angles required to identify the beginning point.

The process is illustrated diagramatically in Fig. 6.3d

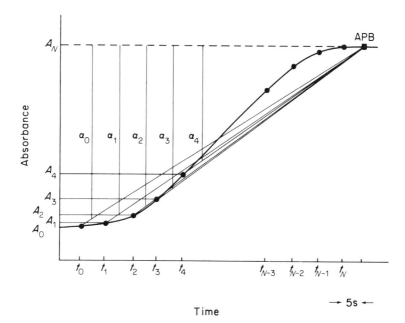

Fig. 6.3d.   *Determination of the start of linearity*

— A time point five seconds beyond the final reading is chosen for the APB.

— Lines are drawn to each absorbance value obtained. The angles are then compared in turn beginning with that obtained from the first time point (α_0).

— When an angle becomes less than or equal to the one previously calculated, then the latter point is regarded as the beginning point of the linear region.

Thus in the profile illustrated.

$$\alpha_0 \; < \; \alpha_1 \; < \; \alpha_2 \; < \; \alpha_3 \; > \; \alpha_4$$

and hence absorbance A_4 is regarded as the start of the linear portion of the profile.

A potential problem with this method is that random spurious results, which might be generated for one of many reasons, would give a false angle reading and trigger the processor into identifying the wrong absorbance value as the start or end of linearity. In order to minimise this problem the processor assesses all the data points and checks that they fall within an allowable degree of scatter. The allowable scatter varies depending on the slope of the reaction line since the steeper this is the more error in measurement is likely to exist and the greater the allowed scatter must be.

The sequence of events is:

— a regression line is fitted to the data;

— the allowable scatter (d) is calculated from:

$$d \; = \; a(m^2 \; + \; 1)^{\frac{1}{2}} \qquad (6.3a)$$

where $a = 0.005$ absorbance units and $m = $ the slope.

— each absorbance point is tested using the relationship:

$$(mt_x + b - d) \leq A_x \leq (mt_x + b + d) \qquad (6.3b)$$

where A_x is the absorbance at time t_x, and b is the intercept on the. y axis.

This identifies the point on the regression line at a given time and compares its absorbance $\pm d$, with the actual absorbance.

A rather different approach to emphasising the start and end of linearity is to calculate certain mathematical derivatives of the original data of which the so-called first and second derivatives are most commonly used. In regions of the reaction profile where the rate is constant (ie the sought-for linear reaction rate region, and also any lag or plateau regions) the calculated derivatives are constant and if plotted produce horizontal lines. Where there is a change in rate (ie at the junction of the lag and plateau regions with the linear region) the derivative values change markedly and make this position more evident (Fig. 6.3e) The TR analyser from Beckman RIIC Ltd uses this approach.

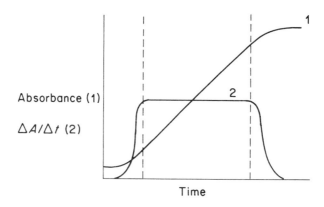

Fig. 6.3e.   *Absorbance–time* (1) *and first differential* (2) *plots for an enzyme reaction*

Π The following table (Fig. 6.3f) contains the absorbance data produced during an enzyme reaction.

Primary data		First derivative		
X (time)	Y (absorbance)	ΔY	ΔX	$\Delta Y/\Delta X$ ($\times 10^3$)
0	0.050			
2	0.050			
4	0.050			
6	0.054			
8	0.060			
10	0.068			
12	0.078			
14	0.089			
16	0.101			
18	0.114			
20	0.128			
22	0.143			
24	0.158			
26	0.173			
28	0.188			
30	0.203			

Fig. 6.3f

(a) Plot these data as a normal graph and attempt to identify the beginning of the region of constant reaction rate.

(b) Calculate the first derivative, plot the data obtained on the same axes as the above and discuss the result.

The data are reproduced, together with the various calculations, in Fig. 6.3g, and the original data and the first derivative are displayed graphically in Fig. 6.3h.

Primary data		First derivative		
X (time)	Y (absorbance)	ΔY	ΔX	$\Delta Y / \Delta X$ ($\times 10^3$)
0	0.050			
		0	2	0
2	0.050			
		0	2	0
4	0.050			
		0.004	2	2.0
6	0.054			
		0.006	2	3.0
8	0.060			
		0.008	2	4.0
10	0.068			
		0.010	2	5.0
12	0.078			
		0.011	2	5.5
14	0.089			
		0.012	2	6.0
16	0.101			
		0.013	2	6.5
18	0.114			
		0.014	2	7.0
20	0.128			
		0.015	2	7.5
22	0.143			
		0.015	2	7.5
24	0.158			
		0.015	2	7.5
26	0.173			
		0.015	2	7.5
28	0.188			
		0.015	2	7.5
30	0.203			

Fig. 6.3g.   *Table showing calculation of first derivative of data given in Fig. 6.3f*

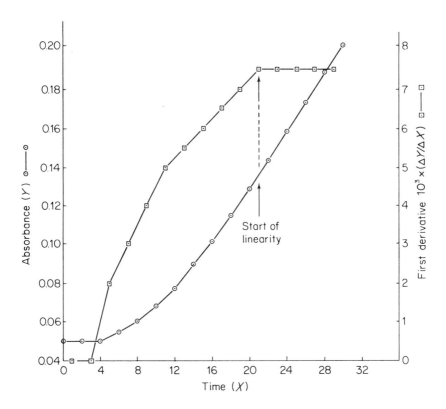

Fig. 6.3h. *The use of first derivative calculations to locate the beginning of linearity*

It is probable that most people would choose 14 or 16 minutes as the beginning of linearity. However the data have been selected to give a very smooth curve in order that the changes are more apparent and even so a careful inspection of the ΔY column in Fig. 6.3g shows that these choices are incorrect. It is unlikely that with a genuine set of experimental data this would be quite so apparent from the tabulated data.

The plotted first derivative shows a sharp change in slope at 21 minutes which would suggest this to be the start of linearity, and the fact that this is the case is confirmed by the Y column in the Fig. 6.3g.

SAQ 6.3a

Define or briefly describe the relevance of each of the following to reaction rate measurements:

(i) signal averaging and elimination;
(ii) measurement of slopes;
(iii) comparison of integral areas;
(iv) scatter limits;
(v) derivative calculations.

Summary

Some practical aspects of the measurement of enzyme reaction rates are described. In particular the relative merits and instrumental approaches to one-point, two-point and multi-point measurements, and the identification of the linear part of the reaction curve are considered.

Objectives

You should now be able to:

- state why enzymes can only really be measured in the kinetic region of the reaction profile;

- define and clarify the nomenclature used to describe the assays undertaken in the kinetic and equilibrium regions of a typical reaction profile;

- state two advantages of assays carried out in the kinetic region;

- explain the principles and assumptions in single-point, two-point and multi-point fixed-time assays;

- explain the differences in practical approach to one and two-point assays between discrete and continuous autoanalysers;

- describe at least three problems or necessary requirements for two-point assays;

- discuss problems that might originate from dilution of enzyme samples;

- describe the advantages of discrete analysers over continuous ones for multi-point enzyme assays;

- explain the basis of at least two approaches to the identification of the linear region in a typical enzyme reaction.

7. Special Problems Concerning the Reliability of Enzyme Assays

Overview

This part of the Unit will emphasise the need for quality control in enzyme assays and the particular problems with the measurement of enzymes compared with non-enzymic sample components. It will also discuss the problems encountered with the preparation and use of specific standards for incorporation into quality control schemes and the desirability of developing standard methods for enzyme assay.

7.1. THE NEED FOR QUALITY CONTROL IN CLINICAL CHEMISTRY

The primary object of these analytical chemistry publications is educational and as such we are mainly concerned with general concepts and principles. If you ever undertake any practical work on enzymes in association with this course, simply obtaining a numerical answer for your experiments would probably be regarded as adequate since biological experiments are fundamentally difficult to perform and it is unrealistic to expect high quality results from first experiments.

However in a practising clinical chemistry laboratory it is extremely important that the numbers generated for the measurement of enzyme activity are meaningful; in other words are as close as is practicable to the actual value found in the sample, or at least show a constant deviation from it.

It is the case that many individuals have a touching faith in numbers and tend to assume that the results generated by experiments are accurate, especially if they are produced by complex procedures or involve apparatus with impressive digital displays. With regard to simple calculators for example many adults let alone school children still need to be acquainted with the important caveat 'garbage in, garbage out'

In order to maximise the likelihood of producing useful experimental results it is necessary to organise a definite, planned scheme of quality assessment as a means of establishing the reliability of the results before they are used for clinical judgements. Most hospital laboratories in the UK take part in such schemes, notwithstanding the fact that it has been conservatively estimated that a comprehensive quality control programme can add 10–20% to the operating costs of a typical Clinical Chemistry laboratory.

In 1975 Buttner defined quality control as 'the study of those errors which are the responsibility of the laboratory, and the procedures used to recognise and minimise them. This study includes all errors arising within the laboratory between the receipt of the specimen and the despatch of the report. On some occasions the responsibility of the laboratory may extend to the collection of the specimen from the patient and the provision of a suitable container'.

This is a far-reaching definition and it is perhaps regrettable that in general, laboratories focus their attention on monitoring the performance characteristics of the analytical methods they employ, whereas simple clerical errors for example, can account for a significant percentage of incorrectly reported results.

Quality control or assurance is in general achieved by the measurement of selected components in specially reserved samples and a comparison of the results obtained with those expected. Schemes for this operate both within laboratories and, on a national scale, between laboratories. So important are these procedures that they are dealt with extensively in the Unit concerning the Assessment and Control of Biochemical Methods.

It is unfortunate that quality control with regard to enzyme assays is much more difficult than for simple inorganic and organic constituents and hence it is relevant to discuss the particular problems involved with enzymes in this Unit. The article by Rosalki (1980) may also be of interest.

7.2. REFERENCE METHODS

7.2.1. The History of Method Development

In the early days of enzyme analyses, laboratories devised their own methods based largely upon considerations of personal experience, availability of reagents and apparatus and also on their suitability for the specific needs of that laboratory. Such methods may not necessarily be 'optimal' in the sense of using the best possible combination of reaction conditions and producing the most reliable results, but nonetheless they might be adequate for clinical needs because the interpretation and clinical evaluation will be made by a comparison of the results of patient samples with known normal ones using the same method.

∏   What do you think are the major problems with such an independent approach?

         It is hoped that the previous parts of this Unit will have made it clear that a more rational approach is necessary for two main reasons.

— In the interests of scientific communication it is important that comparable methods are used.

— An integrated approach between laboratories is useful in reducing the duplication of effort expended in independently developing methods in different laboratories.

Most countries have national professional bodies with interests in these areas who have taken on the role of method development or co-ordination of development within their country. For example:

The United Kingdom Association of Clinical Biochemists,

The Committee for Enzymes of the Scandanavian Society for Clinical Chemistry and Clinical Physiology,

Deutsche Gessellschaft für Klinische Chemie,

The American Association of Clinical Chemistry.

The USA also has a National Committee for Clinical Laboratory Standards.

There is also an international body for clinical chemistry –'The International Federation of Clinical Chemistry' who set up an 'Expert Panel on Nomenclature and Principles of Quality Control in Clinical Chemistry' in 1971 with this as one of its overall terms of reference.

7.2.2. Definitions

With regard to methods, a number of terms have arisen that need definition before we move on.

A *definitive* method is one that has no known source of inaccuracy, ambiguity etc. Such methods have not been devised, and perhaps are in principle impossible to achieve for enzymes, bearing in mind the lability of the reagents and the complexity of the assays involved.

A *reference* method is one with the least inaccuracy of those available and since, by definition, such methods are simply the current 'state of the art' at any given time, they are achievable, and by a process of continual modification should improve with time.

Ideally most reasonably well-equipped and staffed laboratories should be able to carry out assays by the reference method but such methods may not be suitable for routine use due to their complexity, cost, instrument or time requirement etc, and so recommended methods are also proposed.

A *recommended* method is one that careful study has shown to be capable of achieving acceptable standards for the main criteria of an assay namely, accuracy, precision, robustness, sensitivity and adequate detection levels, and is acceptable to laboratories by virtue of its simplicity, cost, time requirement etc.

Another term found in this context is *optimised* method which denotes one in which all the important parameters are adjusted to the value giving maximal activity.

7.2.3. The Determination of Optimised Reference Methods

The determination of optimised methods seems on first thought to be easy, the original set of method instructions being taken and each instruction is then altered in turn, with the value giving the maximal activity noted for incorporation into a revised procedure.

∏   Assuming this is done, can you think of an entirely practical problem that might arise with the implementation of the results of these optimisation studies.

       The values obtained may not be of sizes that are practically useful, eg a substrate concentration of 0.0931 μmol dm^{-3} would be more difficult to prepare than one of 0.1 μmol dm^{-3}. Careful consideration of the effect of deviations from the optimal conditions (ie of the 'robustness' of the method) needs to be made.

There is a fundamental scientific reason why the determination of optimum levels by the method described above may not produce the best combination of experimental variables. Can you think what it is?

Unfortunately the variables may inter-act with each other. So while an optimal coenzyme or substrate concentration of X units might occur at pH 8.0, changes in ionisation state might mean the optimum becomes 0.9 X units at pH 7.0.

Determination of these interactions can be done by the time consuming procedure of re-investigating each variable every time one of them is adjusted to its newly determined optimum level. Thus:

Stage 1: vary parameter A over a range $A1 - A10$, $A4$ proves to be optimal,

Stage 2: vary parameter B using $A4$, $B6$ proves to be optimal,

Stage 3: using $B6$, vary A to check if $A4$ is still optimal for $B6$.

Enormous numbers of combinations and trials are required to investigate fully the interactions even with only a small number of variables.

To help reduce the labour involved mathematical approaches have been employed and the discussions by Krause (1974) and London (1975) provide useful examples of these. One point arising out of London's work that is very relevant to the comments just made is that for the assay they studied, one set of optimal conditions did not exist. In other words a number of combinations of inter-acting conditions gave maximal results.

ie $A4 + B5 + C7$ gave similar results to $A2 + B3 + C6$.

The article by Tietz (1983) includes an interesting discussion of a computerised response surface technique to help visualise the interactions of variables by producing contour plots (Fig. 7.2a). In this case the catalytic activity produced by the combinations of concentrations used in the traditional method which was taken as the starting point for the optimisation study, was quite close to the maximum achievable activity. This was useful since the conditions for the latter required the use of awkward concentrations of the two solutions.

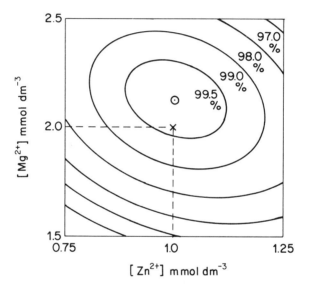

Fig. 7.2a. *Contour plot of the catalytic activity of alkaline phosphatase with different concentrations of magnesium and zinc ions*

⊙ – Combination giving maximum activity,
× – Activity obtained using combination employed in traditional methods.

Other parts of this Unit deal with the importance of isoenzymic forms of enzymes in diagnosis and an unfortunate problem with regard to the development of optimised methods is that the catalytic properties and environmental requirements of the different isoenzymes may not be the same. These variations are made use of in their differential measurement but work against us in the development of optimised methods of analysis.

In the case of AST the mitochondrial isoenzyme has a higher Michaelis constant for 2-oxoglutarate and a lower one for aspartate compared with the cytoplasmic form. Since the proportion of the two forms in the serum will vary from patient to patient, and also with various disease states, there is no single set of optimal conditions suitable for general use.

The various national bodies separately, or collectively via the IFCC, work towards the production of suitable optimised reference methods. These are published as provisional recommendations for open evaluation and criticism before being incorporated into a final version. See IFCC (1976) in the reference list for a specific example.

The IFCC assumes responsibility for a very wide range of aspects of the development of methods and have specifically listed:

(*a*) the clear description of the principles of the measurement;

(*b*) the selection of reaction conditions and the other method variables;

(*c*) the provision of instructions for the preparation of solutions including statements on required purity of reagents and stability of solutions;

(*d*) the standard of instrumentation required;

(*e*) the type of specimen, its removal, preservation, storage etc;

(*f*) the assay method including blank and standard requirements and calculation methods;

(*g*) validation procedures;

(*h*) the publication of provisional recommendations, preliminary studies on the assay, general consultation and ongoing modification.

The recommendations are published in considerable detail (giving pH adjustment temperatures for pH buffers for example); and are certainly sufficient for any laboratory to perform the assay and evaluate it. The recommendations are subject to considerable scrutiny and debate, eg over the choice of pH 7.5 or 7.3 for the ALT assay, or whether the GGT transferase acceptor (glycylglycine) has sufficient buffering capacity to allow omission of another buffer reagent from the reagent mixture. The object of this open discussion is to draw on the experience of many professional laboratories before producing a final version of the method. Full agreement is of course difficult to achieve and reached an interesting development in the early 1980's with the IFCC's Expert Panel on Enzymes recommending 30 °C as the preferred assay temperature, and the Expert Panel on Instrumentation recommending 37 °C.

Notwithstanding these problems, the need for such co-ordination is long overdue, Wilkinson quotes a situation in which a 1976 survey of 300 laboratories carrying out AST assays showed more than 30 method variants to be in use.

The data in Fig. 7.2b are taken from a recent quality survey using Wellcome standard K0838 and they show the differences in target values for several analyses when different reaction conditions are employed. The values are sufficiently different to make a comparison of the results from different laboratories a very uncertain exercise.

(*a*) Variation in Basic Principles of the Method.

Substrate	Buffer solution	Measuring principle	Target value U m^{-3} (For AP)
p-nitrophenyl phosphate	AMP	single or multi-point assay at 37 °C	93
p-nitrophenyl phosphate	AMP	reaction rate assay at 37 °C	79
p-nitrophenyl phosphate	DEA	reaction rate assay at 37 °C	171
phenyl phosphate	*	end-point or single point at 37 °C	88

* Unspecified

(*b*) Variation in Substrate Concentration

Substrate	Measuring principle	Target value U dm^{-3} (For BDH)
3.3 mmolar 2-oxobutyrate	reaction rate assay at 37 °C	722
12–15 mmolar 2-oxobutyrate	reaction rate assay at 37 °C	976

(*c*) Variation in Temperature

Substrate	Measuring principle	Target value U dm^{-3} (For BDH)
3.3 mmolar 2-oxobutyrate	reaction rate assay at 25 °C	489
12–15 mmolar 2-oxobutyrate	reaction rate assay at 25 °C	601

Fig. 7.2b. *The effect of variation in reaction conditions on target reaction rate values for selected enzymes*

These data are taken from a 1984 Wellcome Group Quality Control
Programme using standard lot K0838.

∏   Can you recall from the previous parts of this Unit, why a
      reduction in temperature or substrate concentration should
      give a slower reaction for the assay? ·

      This type of phenomenon is discussed in Sections 1.2 and 2.4
      and it centres upon the fundamental factors affecting the rate
      of chemical and therefore enzymic reactions. An increase
      in reactant concentration will increase reaction rate up to a
      certain limit, by virtue of increasing the number of collisions
      occurring. An increase in temperature will increase the aver-
      age energy level of the molecules and again will produce an
      increase in the number of colliding molecules with sufficient
      energy for the reaction to occur.

Even the use of the methods recommended by the national professional bodies can sometimes only lead to a reduction and not an
elimination of the problem. Fig. 7.2c shows some results from the
UKEQAS programme of early 1984 in which ALT was assayed by
the methods recommended by the Scandanavian (SCE) and West
German (DGKC) Societies. The differences in mean values are significant and especially for LDH there is also a substantial difference
in standard deviation for the two methods

Method source	ALT			LDH		
	Mean result ($U\ dm^{-3}$)	SD	Number of laboratories	Mean result ($U\ dm^{-3}$)	SD	Number of laboratories
SCE	50	6	80	1114	100	49
DGKC	59	7	61	1101	168	65

Fig. 7.2c.   *A comparison of assay results using recommended
methods*

Data are taken from a report on the UKEQAS programme in early 1984. The optimised methods recommended by the Scandanavian (SCE) and West German (DGKC) societies were used on a human serum standard, lot H1QC/9.

SAQ 7.2a	Distinguish between the following terms:
	(*i*) definitive and reference methods;
	(*ii*) optimised and recommended methods.

7.3. STANDARDS

7.3.1. Definitions

An essential requirement for a quality control scheme is the provision of a standard material for regular analysis. In chemical and physical terms, primary standards are materials (in our case generally solutions) of precisely known chemical composition and purity, for example accurately weighed pure substances dissolved in accurately dispensed pure solvents. Secondary standards are materials with a concentration determined by the quantitative analysis of the prepared solution, preferably with reference to a primary standard.

For enzyme work primary standards are impossible to achieve since we are interested in a response (ie activity) rather than a weight of solid or a concentration in a solution. Accurately weighing a solid enzyme preparation is of little use since the enzyme activity per unit mass of solid may vary with different preparations. Another problem is that the determination of absolute activity of an enzyme solution is very difficult for analytical reasons, so even the production of good secondary standards is difficult.

With regard to the numerical value in a standard material a number of terms are encountered which need to be clarified.

— The term *true* value does indeed mean the correct or actual value and it is certainly the aim of analysis to come as close to this as possible.

— The *definitive* value is the one obtained from the definitive method and the *reference* value is the one obtained from the reference method.

— An *assigned* or *certified* value is one given to a material by the manufacturer or by a scientific body; whereas such materials are sometimes said to be arbitrary standards the assignment should be based upon scientific investigations and ought to be anything but arbitrary.

— *Consensus* values are derived from the results of a large number of assays generally as part of a quality control scheme. While the value is only obtained after some delay and is suitable therefore only for retrospective action, it is nevertheless very useful since in a large Quality Control (QC) scheme many results will be used in its determination. Furthermore if a range of methods has been employed and extreme values are excluded, the end product consensus value should be quite close to the correct value.

— *Target* values are mathematically derived from the results of analyses. They represent the range of values for the analysis of a standard material within which a laboratory should reasonably expect to obtain values, if its standard of analysis is adequate.

A distinction needs to be drawn between calibration and control materials, the former are used to calibrate instruments and analytical systems. The material should have an accurately known value and the routine results obtained can be compared with this for the purpose of calculation. Control materials are used for the purpose of ongoing method monitoring, ie as a check on calibration, on the stability of the analysis etc. It is important not to use the values of this material to adjust the raw experimental data from patient's samples by re-calculation in order to allow for any difference obtained between control and calibration results. The observation of a difference should be used as a stimulus to investigate the cause of the difference. It is essential therefore that different materials, ideally with quantitatively different compositions, are used for the two purposes.

7.3.2. The Nature of Standards

The preparation and use of standards for enzyme analysis is particularly troublesome and over the years a number of different materials have been employed.

On first thought it might seem that purified enzymes ought to be used wherever possible but can you imagine what the theoretical and practical problems are with such approach?

— The extensive processing of the original biological extract necessary to generate pure enzymes can produce substantial changes in the nature of the enzyme, for example producing changes in their reaction kinetics.

— In any case it is difficult to produce absolutely pure enzymes since some organic compounds complex with proteins by adsorption, intercalation etc.

— The cost would be rather high due to the labour involved and the wide range of procedures required.

— A significant number of enzymes are less stable in the pure state than in mixtures.

Tissue extracts have also been used but their different composition makes them unsuitable for standardisation of blood analyses. Most commonly sera or plasma are employed. The use of these materials became particularly desirable following the development of multichannel autoanalysers since a serum standard is certainly easier and cheaper to produce than a complex mixture of primary standards even if such a procedure were possible for enzymes. However more importantly the interaction of different serum components with each other is of considerable significance with regard to enzyme activities. Collectively such interactions are called matrix effects and include physical phenomena such as viscosity, surface tension, ionic charge etc, which affect diffusion rates, protein surface charge and hence stability, reactivity and other phenomena. Chemical interactions leading to interference, reduction in reaction specificity, increases in inhibition etc, are well known. Since such effects will exist in the samples to be analysed they should also be present in the standards.

Various types of sera can be used in quality assessments. Individual patient's sera are the most readily available but suffer from the liklihood of being atypical in one or more respects and hence pooled sera are more common. These can be pooled from individuals who are assumed to be normal or from individuals with particular diseases. This latter approach is designed to provide abnormal specimens which would have an obvious value in analyses, but the ramifications of the consequences of even a slight change in metabolism are so extensive that it is difficult to know what the extent of the abnormality in the serum will be. Abnormal sera are more frequently produced by selective supplementation of supposedly normal ones and Fig. 7.3a shows the extent of this for the standard sera produced by one manufacturer.

The Nature of The Standard Control Sera

'Wellcomtrol One' bovine serum with the following additions	'Wellcomtrol Two' equine serum with the following additions	'Wellcomtrol Three' bovine serum with the following additions
–	ALD	–
ALT	–	–
–	AP (human placenta + pig)	AP (human placenta)
AM	AM	–
CK	CK	–
GGT	GGT	–

Fig. 7.3a.   *Wellcome Group Quality Control Programme*

So-called zero reference standards in which selective depletion of materials has been undertaken are occasionally used, especially in determinations of specificity and background reactivity. Production of these used to be quite difficult but the advent of affinity chromatography with its high specificity and efficiency has allowed the selective removal of particular proteins to be undertaken.

An important argument centres around the use of animal sera instead of human in these schemes. The sheer volume of standard material required by all the UK laboratories means that the use of human serum alone would be prohibitively expensive and have noticeable effects on the Blood Transfusion Service. For this reason animal sera are commonly used (eg by the Wellcome Group Quality Programmes), notwithstanding the World Health Assembly's resolution that countries should become self-sufficient in blood products including quality control materials of human origin. While other programmes (eg The UK External Quality Assessment Scheme for General Clinical Chemistry) use human sera they are usually derived from donated blood which has proved unsuitable for transfusion and is therefore by definition, abnormal; they are frequently selectively supplemented by animal enzymes. The reaction characteristics of animal enzymes may of course be different from those of human ones, but the most significant difference is in the isoenzyme pattern. True standardisation of human patient specimens is certainly difficult to achieve since diseases in particular organs frequently cause the release of enzymes with a particular isoenzyme pattern (see Section 8.5), and this is unlikely to be mimiced even in human serum standards, let alone animal supplemented human, or pure animal sera.

A major cause of concern some years ago was the danger to users of the possible contamination of human sera with the vector for hepatitis B, a potentially very dangerous disease easily caught from traces of blood taken in through cuts etc. Nowadays human blood standards are routinely screened for this disease, which is technically easy to do, but adds to the cost. Time will tell what influence the spread of the disease AIDS will have in this area.

Finally sera can be supplied either assayed or unassayed and in the case of enzymes the former will be given an assigned value for the enzyme activities.

∏ In assessing their methods, laboratories are interested in a number of different factors of which accuracy and precision are two of the more important. Which of the two types of sera (assayed or unassayed) can be used in the assessment of both of these, and which of the two factors is the other used for?

  Accuracy of a method describes the closeness of the results obtained to the actual value in the specimen, which presupposes that the value is known or at least an acceptable value for it is available. For this reason only assayed standards can be used to determine accuracy; the unassayed sera are cheaper and are used to monitor the precision (ie replication) of the assay.

7.3.3. The Physical Processing and Storage of Standards for Enzyme Assay

The way in which the prepared serum (normal, depleted or supplemented) is processed by the manufacturer, its storage requirements and stability during storage and use, are of major significance as far as enzymes are concerned because of pronounced changes in enzyme activities that can occur during these treatments.

Three major physical forms of standard have been produced namely liquid sera, frozen sera and lyophilised (freeze-dried) sera; the second requiring thawing and the third requiring reconstitution with water before use. While it may be adequate for a laboratory which prepares its own standards for internal use, to sub-divide the pool and deep freeze the aliquots, transport problems mean that commercial manufacturers generally supply lyophilised material. Wellcome regard all constituents of their sera to have a stability of at least two years even at 4 °C in the lyophilised state and the UKEQAS gives a conservative estimate of 18 months for most non-enzymic components in its specimens, but will not commit itself regarding enzymes. Lyophilisation has the merit of considerably increasing the stability of enzymic constituents in comparison with mere freezing, as well as being more amenable to large scale production with the benefits of continuity and economy of large batches.

A fairly obvious problem with lyophilised standard material, but one which is frequently treated casually, is that the reconstitution of the powder must be done with great care. Apart from actually adding the correct volume of water, it is necessary to ensure that undissolved powder is not 'blown out' of the vial, sticks to the walls of the vial or is otherwise lost, and that sufficient mixing is employed to allow redissolution to occur. However as far as enzymes are concerned more subtle problems exist. On drying, protein molecules can interact with each other and with non-protein materials in such a way as to change their properties permanently and make complete redissolution difficult. Bearing in mind that the reconstituted sera are frequently turbid (see later), any failure to redissolve fully may not be readily apparent.

A related point is that it is generally assumed by laboratories that the replicate vials supplied by the manufacturers are indeed precise replicates; in other words that the manufacturers quality control is such that deviations in preparation and vial-filling are minimal. Achievement of the former is so difficult especially regarding enzyme constituents that manufacturers avoid the problem by producing batches of material with specific identity codes, and between batches comparison is impracticable. The US National Committee For Clinical Laboratory Standards states that vial filling should be to an accuracy of at least $< 0.5\%$ and preferably $< 0.1\%$. Unfortunately there have been cases in the UK of very poor replication of filling giving rise to humorous comments in the scientific press of dispensation being similar to the filling of cups in a British Rail cafeteria by pouring along the rows from a large pot!

Many studies have been undertaken on the changes occuring following reconstitution of the lyophilised serum, on the stability of the components and their similarity to those components in native serum.

One very common and obvious difference is that the reconstituted sera are usually turbid due to a tendency for the lipoproteins to denature on freezing. This does little to imbue confidence and respect among the laboratory staff, and more importantly makes it difficult to employ the material in blind trials. Other well documented changes are:

— a steady increase in AP activity during the first 18 h following reconstitution and then a much slower one for 2 to 3 days. The net increase is usually 10–20% but can be as high as 50%. It is generally recommended that sera for AP assays are prepared on the day prior to use, but this is not compatible with the needs for other enzymes;

— with regard to the last point, CK and ICDH are relatively unstable after reconstitution and assays should be performed within 8 h;

— reconstituted lyophilised or thawed frozen sera will have a significant reduction of the LDH 4 and 5 isoenzymes due to their instability compared with the other isoenzyme types. In addition the isoenzyme pattern will be altered because of a tendency for the sub-units to dissociate on freezing and reassociate in random combinations on thawing;

— the cytoplasmic isoenzyme of AST is more stable than the mitochondrial one giving rise to the same problem.

In 1982 Jung *et al* published some interesting data on a comparison of the response of various sera to changes in reaction conditions. The enzyme studied was GGT and it was measured by the Scandanavian Society's recommended procedure. A number of method deviations were studied including for example the effect of decreasing the substrate concentration to two-thirds, and also the effect of changing the pH by 0.36 units. In total 11 different commercially available quality control sera were included in the survey but most importantly, their response was compared with that of a normal human serum. The most striking observation was that the QC sera varied substantially in their response to these perturbations in method and in some cases they gave as much as twice the response shown by the normal human serum. Rarely did any of them give a lower response to any of the changes studied.

Recently the Beckman Corporation have introduced liquid QC standards which they call *Decision*. The materials are manufactured from defibrinated fresh frozen human plasma.

∏   What advantages do you think such material will have?

— A major advantage is speed of use since there is no requirement to reconstitute and wait for full re-activation to occur.

— This should mean a reduction in operator error due to the omission of these steps and a reduction in manufacturers bottle to bottle variation.

— Worries over the quality of the water used for the reconstitution cease to exist.

— A marked reduction in waste should occur since reconstituted lyophilised or thawed frozen standards need to
be used within a given time. The manufacturers of liquid
standards usually state that the enzymes should be stable
for at least a year at < 15 °C, and 20 days even at 4 °C.

— The problems referred to earlier over the effects of freezing on proteins (especially the lipoproteins) should disappear, one advantage of which is the lack of turbidity
in these standards at the time of use. The similarity in
appearance to normal patient specimens is regarded as
being of great importance.

On the counter side of the argument is the fact that the standards
are stabilised by high concentrations of polyhydric alcohols, eg 30%
ethylene glycol. The viscosity of the material is more than twice
that of other types of standard material and may be the cause of
the results reported by Elin and Gray (1984) which show that the
coefficients of variation of the analysis of 19 materials (5 enzymic)
were generally greater than for the other standards tested (Fig. 7.3b).
To be fair these results contrast somewhat with those of Tanishima
(1977) which suggest that these alcohols reduce analytical variability
for enzyme assays.

Enzyme	Standards		Standards	
	Decision Level 3	Ortho Abnormal	Decision Level 3	Omega II
AP	3.8	4.6	2.3	2.5
ALT	6.1	3.7	5.5	4.7
AST	4.4	3.0	4.3	4.0
LDH	5.8	3.6	3.0	3.0
CK	10.8	9.6	6.5	6.0

The data are coefficients of variation (%) taken from Elin R and
Gray B (1984). Decision Level 3 is a liquid standard, Ortho Abnormal and Omega II are lyophilised standards.

Fig. 7.3b. *A comparison of analytical variability for three quality
control materials*

SAQ 7.3a

> (*i*) Distinguiush between the terms primary and secondary standards, and assigned and consensus values.
>
> (*ii*) State one advantage and one disadvantage of using animal sera instead of human sera for quality control of enzyme assays.

SAQ 7.3b

> Explain why:
>
> (*i*) reconstituted sera are turbid in appearance;
>
> (*ii*) reconstituted sera are different in their isoenzyme (ISE) pattern compared with normal sera;
>
> (*iii*) supplemented sera are different in their isoenzyme pattern from normal sera.

SAQ 7.3b

SAQ 7.3c

> List at least three problems with lyophilised sera for use as standards in quality control programmes.

Summary

This part of the Unit has been concerned with the reasons why:

— quality control programmes are required in clinical analyses and the particular problems with regard to quality control of enzyme assays;

— the development of standard methods for enzyme assays is important;

— standard reference enzyme preparations are required in quality control, and the problems involved in the production of these.

Objectives

You should now be able to:

● discuss the need for the production of meaningful results from analyses in clinical chemistry laboratories and as a consequence the need for a quality control programme;

● discuss the value in clinical chemistry laboratories of having standard reference methods for enzyme assay;

● state the role of professional bodies in the development of methods and define the common terms used to describe the various types of developed methods;

● discuss the need for standard materials in the organisation of a quality control scheme and define the common terms associated with different standard materials;

● describe the different types of standard material that are available and review their relative merits, including a consideration of the effects of the processing treatments on the enzyme activities in the standard materials.

8. The Basis of Diagnostic Enzymology

Overview

You will already have studied some general aspects of enzymology, in particular the chemical nature and biological roles of enzymes, and the ways in which these activities can be measured. This Part of the Unit will show you the physiological basis of diagnostic enzymology and the factors leading to its clinical usefulness. It should also indicate the problems that can potentially reduce its value and the factors and technical procedures that minimise or overcome these problems.

8.1. INTRODUCTION

Simple studies on enzyme levels in clinical samples began about 1900, and the subject developed steadily during the first few decades of this century; the demonstration in the 1920's for example, of an increase in phosphatase activity in the blood of patients with certain cancers providing a useful early stimulus. However the subject really took off in 1954 when it was shown that increases in the activity of blood aminotransferases followed shortly after the development of a type of heart attack known as a myocardial infarction. This demonstration was obviously of great interest because of the frequency and seriousness of heart attacks among the population of developed countries.

At the present time it is estimated that perhaps 20% of the entire workload of a typical hospital clinical chemistry laboratory involves the assay of enzymes in patients' samples, for the diagnosis and monitoring of treatment. Hopefully by the end of this part you will understand why this is the case.

NOTE. The health of humans can be affected in many different ways, for example through diseases caused by invading or external organisms ranging from the simplest viruses to large multicellular organisms such as fleas. Other external agents such as drugs, poisons, pollutants etc can have serious effects; mankind is also susceptible to many internal conditions originating from genetic defects, the general breakdown of metabolism, physiology and its control, and the complicated and poorly understood end results of the ageing process.

Unless there is a definite reason for not doing so in the following text we will use the general term 'disease' to cover all of these sources of change merely for the sake of brevity.

8.2. THE BASAL LEVEL OF ENZYMES IN THE BLOOD CIRCULATION

Small quantities of enzymes are continually present in the circulation and diseases can result in changes in the level of these.

From this and the next section you will find out:

— why small quantities of enzymes are present in the blood,

— how changes in the level of these might occur in disease,

— in what directions these changes might occur.

8.2.1. The Origin of Enzymes in the Circulation

It is salutory to note that not only are organisms (including individual humans) not immortal but their components (cells, molecules etc) are even less long-lived. With few exceptions they are continually being degraded and remade; adult human red cells lasting only about 120 days for example and thus being removed at a rate of about 2 million each second. Obviously if the process is even remotely like a bursting balloon, at least some of the contents of these cells will be lost from the tissue or organ into the body fluids including the blood. It does in fact seem that, considering the number of cells involved, remarkably little of the cellular enzyme complement is lost during normal cell replacement.

Perhaps the possibility that enzymes might be lost when cells die is fairly obvious but curiously enough a number of subtle experiments, involving for example radioisotopes as labels to follow the movement of proteins, have shown that enzymes can leak from cells which are not apparently dead or even seem to be dying, and that such leakage can increase dramatically with even mild diseases.

It is important to note that the leakage of proteins across normal cell membranes is likely to be difficult and slow which unfortunately tends to result in a delay before a disease produces measurable changes in blood enzyme levels.

∏   Which of the following properties of a typical enzyme do you think could make leakage difficult?

    (*a*) spherical shape,
    (*b*) construction out of multiple units,
    (*c*) large size,
    (*d*) surface charge,
    (*e*) catalytic activity,
    (*f*) attachment to other structures.

The most significant of these factors is their large size (c), even quite small proteins (most of which are non-enzymic) have a M_r of 60,000 Dalton with many proteins reaching 200 000. Compare that with glucose at 180! Several other factors could be significant also, of which one of the most interesting is the tendency for the cell to organise enzymes from a given metabolic pathway or area into efficient units by entrapping them in organelles, by associating them into particles, or by attaching them in groups onto membranes (f). An important point is that some potentially useful enzymes are so complexed that they are never released unless the cell actually dies, a topic we shall explore later as it has both disadvantages and benefits in clinical enzymology.

We have so far dealt largely with what might be called the accidental leakage of enzymes from cells, occuring when cells die or are damaged – has it occurred to you that there will also be enzymes secreted intentionally into the blood because this is their site of action? While the majority of these are involved in defence mechanisms including blood clotting processes and counter-acting invading bacteria and viruses, some are more directly involved in metabolism (for example the steroid metabolising enzyme lecithin–cholesterol acyl-transferase, LCAT).

8.3. THE EFFECT OF DISEASE ON THE BASAL LEVEL OF CIRCULATING ENZYMES

∏   It is conceivable that a disease could affect one or several of a large number of aspects of cell structure and/or operation, and in this way change the rate of leakage of enzymes into the extracellular fluids. Some of the more important aspects of cell structure are listed below, which of them do you think might induce or increase the release of cellular enzymes if they are adversely affected? A brief indication of their main function is given in case you are unfamiliar with the details of mammalian cell organisation, but you could usefully refer to some of the texts given in the reading list if you would like more information.

(*a*) the endoplasmic reticulum (removal of toxic materials, synthesis of lipids and proteins),

(*b*) the nucleus (storage of the information required to synthesise the correct proteins in the appropriate quantities),

(*c*) the mitochondrion (production of energy in a useful form),

(*d*) the lysosome (digestion of certain absorbed nutrient particles and liquids),

(*e*) the outer membrane (containment of the cell contents and the control of the movement of molecules into and out of cells).

The outer membrane (*e*) is the most significant factor since it has the major roles of actually retaining the cell contents and also, in order to allow the cell to function, in regulating the movement (usually known as transport) of materials in both directions. Even mild diseases seem to affect the cell membranes, visible changes in width and staining properties are seen and changes in physical properties such as charge distribution, electrical resistance etc, can be measured.

You would not be completely wrong if you chose some of the other structures, for example the nucleus (*b*) and endoplasmic reticulum (*a*) because between them they actually synthesise the enzyme proteins, and certainly poisonous antibiotics and hormones are known to affect these structures. Effects on the mitochondrion (*c*) could also be significant since the maintenance of normal membrane activity requires considerable quantities of energy and in its absence the cellular transport mechanisms fail quite quickly. In addition the lack of energy causes the processes ensuring the maintenance of membrane structure to cease, resulting in a slow degradation of the membrane.

Most of these interactions would be slow to affect enzyme release because enzymes are not normally exported and hence a shortage of energy would not immediately affect them, as for example it would affect the movement of sodium ions. Moreover, decreases in the rate of enzyme synthesis would take some time to become apparent since enzymes are already present and functioning in the cell. Significant deterioration of the membrane due to failure in the maintenance mechanisms resulting from energy shortage would also take some hours to appear and would not lead to enzyme loss in the short term. Such changes are important but in the main are only likely to contribute to the more long term manifestations of disease.

Diseases can bring about changes in the level of circulating enzymes in several other ways, one of which is to affect the rate of synthesis. It seems probable that some kinds of liver damage, including reduced bile flow actually result in an increase in synthesis of AP, GGT and some other enzymes, although of course most diseases produce a decrease in synthetic metabolism.

Another more long term effect of some diseases is to produce a change in the number of cells, Many conditions (alcoholism for example), produce a liver condition known as cirrhosis where large numbers of cells die resulting in a reduction in the total number of molecules available for release, and of course many cancers result in substantial increases in cell number as implied by the popular term 'growths'. One factor that in general does not seem to influence the basal level is previous illness or disease. Assuming the individual makes a full recovery as shown by the usual medical indicators, the level of enzyme in his circulation should return to within the normal limits. This is of major significance in the diagnosis of subsequent diseases since it is therefore still possible to do it by a comparison with the accepted population normal ranges.

8.4.   FACTORS AFFECTING THE USEFULNESS OF ENZYME MEASUREMENTS IN CLINICAL STUDIES

∏   In advance of our continuing discussion you might like to bear in mind the existence of a normal level of circulating enzyme activity, and the fact that diseases cause a change (almost always an increase) in this level, and then ponder on the factors that could affect the usefulness of enzyme measurements. List two or three such factors if you can.

Perhaps the following points are among the most important:

— the magnitude of the basal level for each enzyme;

— variations in this basal level which are not due to disease;

— the size of the increase for different enzymes as a result of disease;

— the severity of the disease before an increase can be measured;

— the time lag from the development of a disease before an increase can be measured;

— the extent to which those enzymes which can in fact be found at increased levels in the blood do actually indicate problems with particular organs (ie their organ specificity);

— the technical ease and reliability of an enzyme assay.

We will now consider each of these factors in more detail.

8.4.1.   Variation in the Basal Level Among Different Enzymes

Detailed discussions of this point are given in many texts including Varley (1980), Brown (1982) and Martin (1978), but it is noteworthy that basal level ranges vary between extremes such as (GPD), which should be at zero in the normal individual, to (CHE) at 0.6–1.4 U dm^{-3}, (AST) at 5–15 U dm^{-3}, and to as high as 240–525 U dm^{-3} for LDH.

From an analytical point of view the enzyme assay is likely to be more reliable if the disease produces a high *percentage* rise rather than just a numerically high one. The lower the basal level the greater this percentage rise is likely to be and it is one reason why assays of GPD are very useful in several investigations. It is apparent from the figures just quoted that the basal level varies significantly from enzyme to enzyme.

Another very important point is that the values are given as ranges. The existence of such ranges of values rather than single constant values, significantly affects the ease with which decisions can be made from single or even a short series of enzyme measurements on patients' samples. The variation implied in such quoted ranges occurs both within a given individual from time to time, and between normal individuals in a population.

Matters are made a little more complex by the fact that whereas the distributions may be Gaussian for some enzymes eg the population range for CK; for others (eg the AST, ALT and GGT transaminase population ranges), they are skewed. The use of some simple statistics in quality control and for other purposes can become quite difficult therefore.

8.4.2 The Influence of Non-disease Factors on Basal Level

Among the non-disease factors affecting reference ranges for the basal level are:

(*a*) assay technique;
(*b*) sex;
(*c*) age;
(*d*) body weight;
(*e*) physiological state.

Let us develop these points in turn.

(*a*) Problems originating from the assay techniques include:

— sampling methods (AST levels increase by about 10% if a tourniquet is used);

— sample storage (SDH levels fall by 22% per 24 h at 4 °C, whereas AM is stable for 7 days even at room temperature);

— direction of measurement chosen for those enzymes with an equilibrium constant near enough unity for the reaction to proceed relatively easily in either direction and hence provide a choice of measurement direction;

— measurement by reaction rate techniques rather than equilibrium ones.

Some of these points are particularly significant for LDH and have generated published ranges varying between 24–78 and 240–525 U dm^{-3}, for assays on normal samples. This obviously creates problems in the investigation of disease states and in particular, emphasises the need for quality control and consistent approach in enzyme measurement.

(*b*) sex (ie gender). GGT varies from 4–18 U dm^{-3} in females to 6–28 U dm^{-3} in males.

(*c*) age. As with many other physiological features children and old people can be markedly different from the rest of the population, the former due to their different physiological make-up and the latter due to a tendency to produce large numbers of defective (and therefore inactive and undetected) molecules. The data for AP are illuminating with adults ranging from 4.8–13.5 U dm^{-3} and children from 8.8–21.2 U dm^{-3}.

The changes with age can unfortunately be quite complex and seem-ingly irrational. Alkaline phosphatase (ALP) rises during puberty and then rises again dramatically after menopause; conversely AST falls progressively with age to a minimum at 30 years and then rises (but only in females). CK is curious in falling in males but rising in females.

(*d*) body weight. This will plainly affect enzymes (eg CK, LDH and AST) originating from organs with a major contribution to bodymass, eg muscle and fat tissue, and the three examples quoted all have high levels in proportion to the individual's body size.

(*e*) physiological state. This term encompasses a wide range of fac-tors such as:

— meals (AM and ALP rise after meals);

— malnutrition (CHE decreases due to reduced synthesis, ICDH rises due to liver damage);

— drug therapy (phenobarbitone increases GGT synthesised by the liver);

— rhythms (diurnal changes affect AP activity during the day);

— pregnancy affects many enzymes (CHE declines in activity);

— muscular activity (cell damage induces release of LDH, CK and ALT);

— posture (when supine, fluid drains back into the blood from the tissues giving a dilution effect; but it may also stimulate metabolic changes of various kinds via osmotic effects. AST declines by 10–15% when supine).

∏   Now here is an apparently fairly complex question concern-ing the use of reference ranges. Take your time over it and do not look for complex answers; it is all fairly straightforward.

A junior technician reports that a serum enzyme activity for a particular patient who was admitted the previous day is 50 U dm^{-3}, and excitedly tells you that the laboratory's reference ranges were redetermined only recently on a new reaction rate analyser, using samples taken from patients in a gynaecology ward plus volunteers among the visiting fathers. Since the reference values obtained were 5–15 U dm^{-3} for females and 8–25 for males, then the patient must obviously be ill.

However, you check and find that your patient is a young, male coalminer who fainted while at work and injured his leg when falling on machinery; and that additional reference range data are available. Which of the following comparative pairs of reference ranges might be of value in aiding in diagnosis?

(i)   pregnant versus non-pregnant women;

(ii)   0600 h versus 1500 h samples from the gynaecology ward;

(iii)   school clinic surveys versus paediatric ward patients;

(iv)   athletes from a major games meeting versus office workers;

(v)   surveys from an old peoples home versus a group of mixed staff from your laboratory;

(vi)   the school clinic survey in (iii) versus the laboratory staff in (v);

(vii)   the recently abandoned manual assay technique versus your new reaction rate analyser.

Bearing in mind that the two reference ranges originally quoted by the technician are from rather specialised populations (pregnant, supine women, and young, mobile males) it is possible that the ranges may not match this patient very well.

Since he is also young and male, population ranges in categories (*i*), (*v*), and (*vi*) are unlikely to be more useful, and since his assay was also done using the new analyser, (*vii*) would be of little interest. However, ward procedures might have resulted in an early collection whereas the male visitors would probably have been sampled during the working day so if the enzyme was one which has diurnal variations in activity this might become apparent from comparison (*ii*) above. Survey (*iii*) might also be interesting since it was done on two populations of young people and might show changes due to posture and enforced rest.

However, range (*iv*) is likely to be most important since his high levels could be due to his greater muscle mass and the strenuous activity resulting from his work as a coal miner; or from the physical damage done to his muscle cells by the accident.

While this is of course a hypothetical example taken rather to extremes, the enzymes LDH and CK would indeed show elevations in such an individual above those found in the bulk of the population.

8.4.3. The Size of the Elevation Above Basal Level

The material presented above might suggest that the value of enzymes in clinical studies could be very limited, however a significant point is that even in normal individuals the ratio of enzyme activity within cells to the basal circulating level for most enzymes is between $10^2 : 1$ and $10^4 : 1$. As a consequence of this releases of quite substantial quantities of enzyme can occur as a result of disease which makes the fluctuations in basal level less significant and the effects of mild cell damage frequently apparent.

In 1958, in a lecture to a German Congress, Bucher made the following statement which quite aptly sums up the situation; 'Considering the very great differences in enzyme concentration between the intra- and extracellular space, and the extraordinarily large surface of cells compared with the space occupied by the plasma, it is an astonishing phenomenon that the plasma is so poor in enzymes'. It is surely quite an effective tribute to the quality of cell membranes and the metabolic salvage processes.

The following facts further illustrate the point:

— a myocardial infarction (heart attack) frequently produces a 7 fold rise in LDH and a 20 fold rise in AST activities in the blood;

— infective hepatitis can increase AST activities by as large a factor as 250 fold;

— as mentioned earlier virtually any rise in GPD activity is likely to be significant since the normal basal level is around zero.

A number of pathophysiological factors affect the size of the elevation seen in practice and it is important that we study these in some detail.

Perhaps the most readily apparent factor is the severity of the damage done, it being fairly logical that the more substantial the damage, the more cellular enzyme is likely to be lost. However this situation is more complex in that the type of damage done and the rapidity of its onset are closely involved with it. Diseases that result in cell death are likely to cause the release of more enzyme than those that do not. Among the latter, diseases that only result in sub-cellular changes may produce very limited serum effects indeed, especially compared with those affecting the cell membrane.

The ease of loss of enzyme from the cells is also significant, with cytoplasmic enzymes being in general more easily released than say mitochondrial ones. Such differences are particularly apparent for diseases resulting in only mild cell damage. Other factors could be important in this context, eg molecular size, surface charge, and association into large multi-enzyme complexes, may well influence the rate of movement across intact cell membranes.

The rapidity of onset is an important point since rapidly developing (ie acute) conditions can result in the rapid release of substantial amounts of enzyme which are rapidly apparent in the circulation since they swamp the normal mechanisms for the removal of these molecules. More slowly developing (ie chronic) conditions may, given time, release the same quantity of enzyme but if the molecules are readily removed from the circulation, a distinct elevation may not be noticeable. We shall return to this point later because the precise pattern of enzymes released can give important clues to the severity and nature of the damage, which is of great importance in diagnosis.

In some cases the time since the condition developed is of significance and this is especially true of very rapidly developing conditions such as certain types of necrosis. In the case of a heart attack, an area of heart muscle may die quite rapidly and the release of cellular enzymes would then occur over a relatively short period. Bearing in mind that the blood contains systems for the breakdown and removal of these enzyme molecules, the size of the elevation depends upon the time delay after the event until the laboratory assay is undertaken.

A point arising out of the above discussion is that the rate at which the released enzymes are removed is of importance particularly since it seems to vary substantially with different enzymes. For lipases the half-life is 3–6 hours, for LDH 4–5 days, whereas for CHE it can be as long as 10 days. A topic we shall develop later is that some enzymes have a number of structurally different forms called isoenzymes. In some cases even these closely related molecules can vary in their half-life; LDH type 1 has a long half-life and the type 5 a short one for example. Variations in the time delay before sample assay may well therefore affect the isoenzyme pattern seen in serum samples.

∏ Consider the following experimental observations, and discuss what they mean regarding the nature of the elimination mechanism for circulating enzymes.

(*a*) Removal of the kidney or liver from healthy experimental animals does not produce an increase in enzymic half-life.

(*b*) Injection of purified enzyme into the vena cava or hepatic portal veins, both of which feed into the liver, does not produce a change in half-life.

(*c*) If purified enzyme is injected into the general circulation the rate of fall in enzyme activity is very much greater than the fall in immunological activity.

(*d*) Similarly if ^{125}I-labelled enzyme is injected, the rate of fall in enzyme activity is much greater than the rate of fall in radioactivity.

All of these experiments tend to show that inactivation and destruction occur in the circulation itself, rather than within the general tissues of the body or in certain major organs.

Observation (*a*) shows the enzyme not to be eliminated in the urine and (*a*) and (*b*) show it not to be destroyed in the liver. The loss of activity of injected enzyme could be due to a variety of inhibitory or denaturation effects, but loss of activity as a consequence of degradation would lead to a retention of radio- and immunological activity for some time until the inactive fragments were removed from the circulation.

∏   Take a hypothetical enzyme with a normal basal level of 20 U dm^{-3}. As a result of a disease the level rises over a very short period to 220 U dm^{-3}.

If this enzyme had a half-life of 5 hours what would be the activity in a sample removed and analysed the following day, say 20 h after the event?

What would be the activity if the half-life was 10 h?

With a half life of 5 h, a 20 h delay would result in the lapse of 4 half-lives and only 6.25% of the original activity would remain (100 → 50 → 25 → 12.5 → 6.25). This is 12.5 U dm^{-3}.

With a half-life of 10 h, only 2 half-lives would have passed leaving 25% of the original activity (50 U dm^{-3}).

I hope you are beginning to see how difficult the interpretation of the results of enzyme assays can be, since:

— it is rare for an instantaneous release of enzyme to occur, more commonly a slow release occurs,

— the normal basal level (20 U dm^{-3} in this example) will actually be a range, say 15–35 U dm^{-3},

— the extent of elevation will depend on the extent of damage and this will affect the 220 U dm^{-3} value,

— the half-life value will also be in a range,

— there are still more factors involved as we shall see in the continuing discussion.

Another variable is the ease of passage of the enzyme from its site of cellular release into the general circulation, from which samples are taken, before degradation occurs. As mentioned earlier the former is influenced partly by the extent and nature of damage done but there are other important points. The absence of a basement membrane in the liver makes loss to the blood system relatively easy compared with many other organs. In this context the blood flow to the organ is of some significance since samples are rarely taken from the capillaries and larger vessels within the organs, and molecules which are removed slowly from the organ into the general circulation are more likely to be degraded before assay. A lower activity will then be apparent.

The volume of the 'sink' into which the released enzymes are distributed can sometimes be important since some low molecular mass enzymes (eg amylase at 45 000 Dalton) can pass into the tissue fluids from the blood system and thus become more extensively diluted compared with larger molecules trapped in the blood vessels. Amylase is also small enough to pass into the kidney tubules and be excreted in the urine which would again reduce the serum activity.

The size of the organ affected by the disease can be significant especially if the whole organ is affected, since there will be more damaged cells present to release the enzyme. Such size differences can be dramatic with a typical adult human liver weighing 1.4 kg, whereas the thymus reaches a maximum weight of only 40 g at puberty. A related point is the number of enzyme molecules actually present within each cell, and this can vary over many orders of magnitude for different enzymes. The greatest organ specificity may well come from the more uncommon metabolic pathways, but these, and their associated enzymes, are likely to be present at a low activity within the cells of a particular organ compared with those of a quantitatively very important process such as protein synthesis. A combination of low cell enzyme activity and small organ size could result in a very small release into the circulation.

∏   Out of each of the following pairs, which factor will tend to give the greater increase in serum enzyme activity. You should try to justify your choice.

(*a*) (*i*) The development of a cancer resulting in an increase in cell number compared with (*ii*) one that does not.

(*b*) (*i*) The development of a relatively mild condition resulting in cellular damage compared with (*ii*) one resulting in cell death.

(*c*) The release of two enzymes, (*i*) one with a half-life of 45 h and (*ii*) the other with a half-life of 36 h.

(*d*) (*i*) The development of substantial oedema (accumulation of tissue fluid) compared with (*ii*) conditions that are non-oedematous.

(*e*) (*i*) The development of a condition in one kidney (weight 120 g) compared with (*ii*) a condition in the liver (weight 1.4 kg). (There are several points to the answer here).

(*f*) (*i*) effects on the membranes of the lysosomes compared with (*ii*) effects of poisons on the ribosomes. (Lysosomes and ribosomes are intracellular structures; lysosomes contain very powerful digestive enzymes, and ribosomes carry out protein synthesis).

(*g*) Effects on the liver resulting in the release of (*i*) a very low mass enzyme (say 40 000 D) and (*ii*) a high mass enzyme (250 000 D)

Detailed discussions of these points are given in the text and while the following brief notes should help to refresh your memory, bear in mind that in most cases the situation is likely to be more complex.

(*a*) (*i*) More cells means more enzyme molecules available for release.

(*b*) (*ii*) cell death is likely to release more of the cell contents.

(*c*) (*i*) For an equal rate of release the longer the half-life the longer the released molecules will last, thus contributing to a higher overall value for the serum activity.

(*d*) (*ii*) Oedema increases the tissue fluid volume and will result in more extensive dilution of those molecules small enough to pass into it. The causes of oedema may result in or from changes in the capillary walls allowing more molecules or larger molecules to pass out of the circulation. Oedema therefore results in a decrease in serum activity for some enzymes.

(*e*) (*ii*) the liver being the larger organ should release more enzyme for a given amount of cellular damage. Furthermore the kidney disease may result in an impairment of the 'blood sieving' mechanism resulting in a loss of enzyme molecules to the urine and an even lower enzyme activity.

(*f*) (*i*) In the long term a reduction or cessation of protein synthesis will result in fewer molecules available for release. However an adverse effect on the lysosome may well release its contents which, as very active degradative enzymes, are powerful enough to kill the cells and cause substantial release of cell enzymes into the circulation.

(*g*) (*ii*) The low mass enzyme is likely to be lost into the urine and tissue fluids and would not therefore tend to give as high a serum activity as the high mass enzyme.

SAQ 8.4a

(*i*) A large group of enzymes present in blood comprise those that are released from cells in various 'accidental' ways. Other enzymes in blood have definite functions there – state two distinct such functions.

(*ii*) State two changes that can be experimentally demonstrated to occur in cell membranes as a result of diseases.

(*iii*) Would an effect of a disease on DNA structure (and hence on the structure and production rate of enzymes) produce earlier or later changes in circulating enzyme level than an effect on the outer cell membrane?

(*iv*) Select from the following list the enzyme which is likely to be most easily released from cells as a result of disease.

(*a*) cell surface lipase;
(*b*) mitochondrial ATPase;
(*c*) cytoplasmic aldolase;
(*d*) Golgi body glycosyl transferase;
(*e*) nuclear RNA polymerase. $\longrightarrow$

SAQ 8.4a
(cont.)

(*v*)   Can you give two examples of different ways in which the assay technique for an enzyme can affect its reference range.

(*vi*)   Choose one of the enzymes from the following list which has a normal blood activity showing a relationship to body weight of the individual.

LDH, AP, GGT, AM, CK, GOT, G-6-PD.

(*vii*) List with examples three physiological states that can alter a patient's enzyme activity from the reference range of the whole population in the absence of disease.

(*viii*) On average which of the following ratios could one expect to find for the cellular: blood activities of a typical enzyme?

$10^6 : 1, \ 10^5 : 1, \ 10^3 : 1, \ 10^2 : 1, \ 10^{-2} : 1$

(*ix*)   Diseases can result in a change in the basal level of an enzyme by increasing the leakage from damaged or dying cells. State two other distinct ways in which diseases can change basal levels.

SAQ 8.4b

The myocardial infarction (MI) is a condition that develops very quickly, whereas many types of liver hepatitis are due to infections and develop more slowly. The enzyme profile in serum following an MI is very similar to that of the heart muscle whereas the correlation is much poorer for hepatitis. What might be the cause of this difference?

SAQ 8.4c

An enzyme is released from a damaged organ at a rate of 100 units per day but it is degraded at a rate of 10% of its activity per day. For simplicity we will assume the processes occur stepwise (which of course is not the case at all), and the following changes will then be found.

Time (days)	Expected activity (100 U day^{-1} released)	Actual observed activity (10% degraded per day)	Actual observed activity (30% degraded per day)
1	100	100–10 = 90	
2	200	190–19 = 171	
3	300	271–27 = 244	
4	400	344–34 = 310	
5	500	410–41 = 369	
6	600	469–47 = 422	
7	700	522–52 = 470	

Thus the level after one week will be 470 units rather than 700. Carry out the same calculation assuming a 30% degradation rate and comment on the result.

8.5. ORGAN SPECIFICITY

In order to use enzyme assays as a means of locating the particular organ that is affected by the condition it is necessary to know which, if any, of the available enzymes are specific to particular organs.

Most cell types carry out the really fundamental and essential metabolic processes by using a common set of metabolic pathways and as a consequence the enzymes of these pathways cannot be specific to one cell type. Enzymes of the respiratory pathways, protein synthesis, fatty acid oxidation etc are of this sort. Even if such enzymes are released into the circulation as a result of disease, a measured increase will only tell you that something is wrong, it will not tell you in which organ/tissue/cell type the disease lies. While this principle is in general true, there are a some occasions when such enzymes can be usefully studied, in particular if the cells can be isolated before assay rather than blood assays being carried out for released enzymes. Thus the reduction in pyruvate kinase (PK) levels seen in isolated red cells is an important indicator of the disease hereditary non-spherocytic haemolysis.

On the whole it is necessary to find areas of metabolism which are characteristic of particular cell types and of course many such areas exist because these are the biochemical characteristics that give the specialised functions to the different organs and tissues. So the urea cycle enzymes of the liver, the digestive enzymes of the gut, and the steroid synthesising enzymes of various endocrine glands are specific to these cells. However in many cases these enzymes are not of any real use in diagnosis either because they tend not to appear with a significant elevation in the blood or because, their assay is difficult, expensive or unreliable. Some highly specific systems are known to be of value, eg the relationship between the increase in serum AM activity and gut conditions such as pancreatitis or ulcers, but unfortunately such instances are relatively uncommon.

Accordingly organ specificity must be improved if at all possible and a number of procedures are available for this.

One approach that is well worthy of mention but can be dispensed with quickly, is to couple enzyme measurements with other investigations such as the analysis of metabolites (urea, cholesterol etc) and even with non-chemical investigations such as ECG studies. If however we restrict ourselves to a consideration of enzymes, then there are two main approaches; one being to measure the so – called isoenzymic forms of enzymes and the other being to make use of enzymes with an asymmetry of distribution which is not absolute. In this case it is usual to increase the number of enzymes measured in order to determine enzyme ratios and patterns.

8.5.1. Isoenzymes

The 1950's saw a considerable improvement in techniques available for the routine separation of proteins and, as a consequence, knowledge of the nature of body chemicals increased dramatically. In particular it was shown that many enzymes are present in a number of forms, and since these carry out the same function they have been termed isoenzymes (ISEs). They differ in a number of features such as the mode of their regulation, their reaction kinetics and so on, and it is the case that these ISEs are frequently distributed unevenly across the various cells of the body due to the different physiological needs of these cells. In some cases the ISEs are unevenly distributed within cells; AST for example is present in different forms in the cell cytoplasm and mitochondria. About 10% of sub-cellular enzymes that have been studied closely have proved to have isoenzymic forms although most of these have not as yet proved diagnostically useful.

With the ISEs being structurally similar and carrying out the same reaction it could possibly have been the case that their differentiation and measurement would prove to be difficult. Happily this has not in general proved so and two main approaches have been used to investigate the ISEs of importance in clinical chemistry.

In one of these the ISEs are sufficiently different structurally to be amenable to separation by chromatography or more usually by electrophoresis. The separated ISEs are then measured by an assay making use of their catalytic ability. The alternative approach is to use differences between the ISEs in selected properties, such as their sensitivity to inhibitors (including antibodies), temperature, pH or their ability to use alternative substrates.

Probably the best example of the use of ISEs is the enzyme LDH, and as we shall discuss this in some detail later, we could instead usefully take AP as a brief illustration of the general principle at this point in the text.

The blood level of AP is raised in a number of conditions, eg prostate carcinoma or damage, haemolysis, Gaucher's disease (a lipid synthesis disorder), bone cancer and Paget's disease (a bone disorder). However the ISE present in the prostate is specifically inhibited by L-tartrate and the difference between blood assays with and without L-tartrate gives a useful, specific measurement of this ISE. An indication of the presence of this potentially serious cancer can thus be obtained.

∏   If AP levels are to be measured it is considered wise to remove the blood sample before carrying out any rectal examination. Bearing in mind the anatomy of this part of the body, why do you think this procedure is recommended?

Remembering that the prostate glands are situated in the groin and that the activities of AP in cells of the prostate are 100–400 times greater than those of any other tissue, the slight damage that might occur by bruising etc during rectal examination is sufficient to raise the serum activity noticeably. Blood samples should therefore be taken in advance of the examination in anticipation of this release.

8.5.2. Enzyme Patterns

There are perhaps two main concepts involved in this approach to improving the organ specificity of enzyme assays.

(*a*) One of these is that instead of attempting to achieve organ specificity by the use of just a single enzyme a number of enzymes could be employed. Thus if the normal level for a series of 3 enzymes is 100 U dm^{-3} each, and each of the 3 become raised to 110 U dm^{-3}; while individually these values may be within the population range, the fact that 3 different enzymes all show an elevation could be significant. The situation is perhaps similar to the legal one of circumstantial evidence; each piece may not be sufficient to convict but collectively a number of pieces might be enough for a conviction.

(*b*) Of greater importance is that by integrating the information obtained from all of the enzyme assays more information can frequently be obtained than from enzymes individually.

The following example is based upon that given by Schmidt and Schmidt (1975) and for it you should refer to the normal serum ranges and organ concentrations for some important enzymes given in Fig. 8.5a.

Enzyme	Organ activity U g^{-1}				Serum activity U dm^{-3}
	liver	heart	muscle	red cells	
AST	96	52	36	0.8	5–20
ALT	60	3	3	0.1	5–25
CK	0.7	350	2030	<0.01	0–50
GLDH	60	1	0.5	<0.01	0–1
LDH	156	124	147	36	80–240

Fig. 8.5a. *The distribution of some important enzymes*

While a serum AST activity of 50 U dm^{-3} would indicate that one of the major organs has been affected by disease, it would not, on its own, allow a reliable distinction between say a liver and a heart condition. If the ALT value is low (AST/ALT ratio > 1) then one is still uncertain as to the origin of the two enzymes. A high serum ALT level (AST/ALT about 1) is however strongly suggestive of a hepatobiliary disease due to the high concentration of ALT in the liver. Such a diagnosis could be achieved by an assay of ALT alone however. Similarly if the CK activity is high a muscle problem of some sort as indicated. An inspection of the table suggests that if the level is very high then skeletal muscle should be the most likely candidate but if the level is only moderately raised then the heart would be more likely.

However remember some of the factors that affect the size of the elevation in serum activity; for example the extent of damage caused by the disease. A serious heart problem could give a higher serum value than a mild limb muscle one. It is here that a pattern investigation is useful. Since all cells contain a large number of enzymes which, within limits, are present in particular ratios to each other, it is the case that at least some of them are released 'in parallel', ie at the same rate. Thus a CK/AST ratio of say 5 suggests the heart is the source of the enzymes whereas a ratio of > 20 suggests skeletal muscle to be important. In many laboratories a ratio of 9 is taken as dividing the two groups.

An LDH level of say 150 U dm^{-3} would tend to exclude the liver, muscle and heart as sites of disease but then the other enzyme assays do that also. A raised LDH value would suggest any or all of these three to be the site but an examination of which of AST, ALT and CK are raised, enables the origin to be narrowed down. Raised LDH and ALT suggests a liver origin, and a raised LDH with AST and CK a heart one. The value of the collective pattern is further confirmed since if the other enzymes (AST,ALT and CK) prove to be in their normal range then the most likely origin of the LDH would be the red blood cells in cases of anaemia or haemolysis.

Patterning can give information on other important aspects of disease, eg:

— a comparison of long half-life and short half-life enzymes indicates whether a disease is acute (ie rapidly developing) or chronic (more slowly developing). By definition in a chronic condition that has existed for some time, enzymes will have been released for a long period and the shorter half-life types will tend to have been degraded. Thus the ratio is quite high compared with the more acute conditions, ALT has a half-life of about 48 h whereas AST has a half-life of about 18 h; consequently they provide an ideal ratio for such investigations,

— the ratio of cytoplasmic and mitochondrial enzymes (c/m) can act as a useful indicator of the extent of cell damage.

Π   What would you expect to find happening to the c/m ratio as cell damage becomes more severe?

It is likely that with mild cell damage only cytoplasmic enzymes would be released; but as damage became more severe, organelle based enzymes would be released resulting in a decrease in the c/m ratio. The aminotransferases again provide a useful example with all of the ALT being cytoplasmic, but only 60% of the AST being cytoplasmic, the rest being located in the mitochondria.

A major importance of enzyme patterning is the ability of it to localise the site of damage or type of damage within a complex organ such as the liver. Differentiation of many of the common liver diseases can be done by a comparison of the serum activities of a number of enzymes principally the aminotransferases AST, ALT and GGT, and also CHE. Since the liver is metabolically a complex organ a wide range of defects are possible and hence differential diagnosis is an involved, yet important matter. The subject is dealt with in detail in Schmidt and Schmidt (1976) and later in this Unit (Part 9), so a single example will be sufficient to illustrate the point here.

The normal liver contains about 95 U of AST g^{-1}, and 60 U of ALT g^{-1} giving a AST/ALT (or 'de Ritis') ratio of about 1.6. Measurement of the ratio in serum can produce three extreme situations ie ratios of 1–2, 0.5–1.0, or of < 0.5. While these ratios indicate increasingly less severe cellular damage they are only of marginal value in precise diagnosis since each can be produced by a range of conditions. A ratio of 0.7 can be generated by chronic hepatitis, fatty liver, poisoning and jaundice due to bile duct blockage, as well as by some other less common conditions. Further differentiation requires the study of additional enzymes, for example ALP is useful in identifying cholestasis (bile duct blockage). The normal serum level of ALP is 20–105 U dm^{-3}, a slight to two fold elevation suggests hepatitis, a 3 to 10 fold rise suggests obstructive jaundice or poisoning. The former is more likely if GLDH levels rise to 25–50 times normal, and GLDH is a particularly useful enzyme since its basal level is very low (Fig 8.5a).

∏   Measurements of ALP and GLDH are generally considered to be good indicators of bile system malfunction particularly cholestasis; whereas ALT, and GGT tend to suggest damage to the liver parenchymal cells. In some cases of severe and/or prolonged cholestasis however, liver cell enzymes can show a significant rise. This illustrates a very important principle affecting diagnostic enzymology, what do you think it is?

It could be that the bile system disease has resulted in damage to the liver cells also. In all our discussions of organ specificity we have tended to assume that only one organ, cell type or even sub-cellular component is involved; unfortunately in many cases this is not true and then very complex enzyme profiles can be generated making diagnosis difficult. In this particular clinical condition the bile released into the circulation as a consequence of cholestasis can in itself cause damage to the liver cells.

Another useful example of the discrimination of affected site within an organ is the differentiation of right and left side ventricular failure of the heart. A typical myocardial infarction gives a AST/ALT ratio of > 1. If the heart is affected by other diseases which cause a left-sided ventricular failure then, assuming no other complications exist, very little change in normal serum enzyme activities are seen. Right-sided failure on the other hand, markedly affects both GLDH (which rises 20–100 fold) and the aminotransferases, probably by a parallel effect on the liver. AST/ALT ratios of < 1 are found due to the release of ALT from the affected liver.

It is hoped that this discussion has convinced you of the usefulness of enzyme pattern in diagnosis. In the knowledge that such patterns are of value, hospital laboratories do not measure the enzymes sequentially with a decision as to the next enzyme to be investigated being based upon the result just obtained. Instead a profile approach is undertaken, in which a particular set of assays is performed based upon the suspected area of disease (liver, muscle, kidney etc). Modern sophisticated data acquisition and manipulation equipment is capable of integrating the data from many such assays and displaying and highlighting sets of data which are outside the appropriate reference ranges. The techniques in question are called 'cluster analysis' and a crude illustration of the sort of visual display that can be produced by these systems is shown in Fig. 8.5b. Occasionally these multi-assay profile approaches result in a waste of resources and effort when unwanted data are produced but in general they are considered worthwhile.

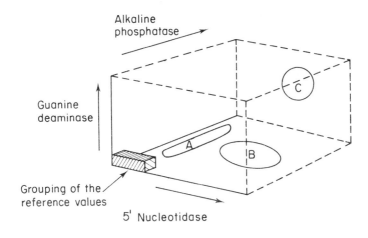

Fig. 8.5b. *The visual display of the results of a cluster analysis technique for studying the inter-relationships of three enzymes*

A: elevated alkaline phosphatase (no liver disease)

B: elevated alkaline phosphatase and 5′ nucleotidase
    (obstructive liver disease)

C: elevated alkaline phosphatase and 5′ nucleotidase and
    guanine deaminase (hepatocellular disease)

The visual display system groups the reference range values for the three enzymes into the block in the bottom left hand corner of the display. Elevations of particular enzymes result in a shift along the appropriate axis.

| SAQ 8.5a | In liver cells the AST activity is approximately 60% higher than the ALT activity. However after a period of acute hepatitis the serum activity of AST is much less than that of ALT.

One of the causes of such a difference could well be a differential rate of inactivation of the two enzymes; but bearing in mind the cellular location of the two enzymes, what other explanation is possible? |

| SAQ 8.5b | Some assays on a particular patient gave the following values for the serum enzyme activities.

Enzyme abbreviation	Enzyme activity $U\ dm^{-3}$	Reference range $U\ dm^{-3}$
AST	50	5–20
ALT	20	5–25
CK	15	0–50
LDH	150	80–240

$\longrightarrow$ |

SAQ 8.5b
(cont.)

(*i*) An acid phosphatase level of 500 U dm^{-3} (with a reference range of 10–170), would suggest the possibility of a liver disease. What data from the above table would tell you that the origin of the disease is not in fact the liver?

(*ii*) If the liver were the site of the damage would you expect the AST/GGT ratio to be high or moderate?

(*iii*) What other approach could be made to exclude the liver as the site of disease?

Summary

This Part illustrates the physiological and analytical principles that provide the scientific basis for the use of enzyme assays in diagnosis.

The concept of the existence of a basal level of circulating enzyme activity and the factors affecting its size, the processes involved in the release of enzymes from cells and the effect of disease on the release process are described. Finally there is a discussion of the importance of organ specificity in diagnostic enzymology and the methods available for improving this.

Objectives

You should now be able to:

● write a short general discussion of the value of enzyme measurements in clinical investigations;

● describe the origins of the basal level of enzymes in the blood circulation of normal individuals, identify at least 5 factors affecting the basal level and discuss the concept of normal range in this context;

● list the cell structures affected by disease which could lead to a change in basal level, indicate their relative importance and discuss at least 2 factors which lead to differences in the rates of release of various cellular enzymes;

● describe the factors affecting the extent of elevation seen in the blood in disease states;

● discuss the need for and the problems of relating elevations of basal level of enzymes to the severity of the disease and the time lag from its onset;

● describe and discuss 2 distinct ways of improving the organ specificity of enzyme assays.

9. Specific Examples of the Value of Diagnostic Enzymology

Overview

This Part is the culmination of the whole of the Diagnostic Enzymology Unit because in it we shall undertake a study in detail of some specific examples of the applications of the principles discussed in the previous parts.

This Part is divided into sections which deal with diagnostic enzymology from three different points of view, that is by a study of a selected enzyme, a selected organ and a selected disease.

9.1. A SELECTED ENZYME – LACTATE DEHYDROGENASE (LDH)

We can obtain a useful illustration of the particular importance of some enzymes, and of some of the problems they can generate and the means employed to minimise them, by taking a specific enzyme as an example for further study. LDH is the one that has been chosen for this.

9.1.1.   The Nature of the LDH Isoenzymes and Their Measurement

In 1959, Markert and Moller studied LDH intensively and they, with others, showed it to consist of 5 main isoenzymic forms (ISEs). The individual molecules were tetramers (ie consisted of 4 sub-units), and had different combinations of 2 types of sub-unit; the M type (for muscle origin) and H type (for heart origin).

∏   Write out the sub-unit composition of the 5 LDH ISEs (ie the 5 possible combinations of 4 units out of the 2 alternatives)

The 5 types will have the ratios;

4H(LDH1), 3H1M(LDH2), 2H2M(LDH3), 1H3M(LDH4), 4M(LDH5).

The abbreviations in brackets are the symbols commonly used in Europe to identify the ISE types.

It is useful that the H units carry a more pronounced negative charge since this allows the 5 ISEs to be separated by electrophoresis. Measurement of the individual ISEs can then be accomplished by a catalytic reaction resulting in the deposition of a formazan on the electrophoresis support (3.2.1). Alternatively the enzyme activity can be measured in solution by making use of its catalysis of the lactate/pyruvate interconversion. This however, only measures the total LDH activity and differential measurement of the 5 isoenzymic forms is very difficult. Fortunately, we shall see later that it is sufficient for diagnostic purposes to distinguish the extreme forms (LDH 1 + 2 against 4 + 5) and this can be done more easily. Three main ways have been used, one takes advantage of the fact that the M units are more susceptible to inhibition by urea or heating; and such treatment of a normal serum results in a fall in total activity to 20–40% of the original, this residual activity being due to the remaining LDH 1+2 ISEs. Alternatively LDH 1 + 2 are the only forms able to use the lactate analogue β-hydroxybutyrate (HB) at a reasonable rate, and a normal serum should give a ratio of HB/Lactate activity (ie HBDH/LDH) of 0.6–0.8. However probably the most valuable approach currently available for the quantitation of LDH1 is based on the precipitation of LDH2–5 by using an antibody to the M sub-unit.

9.1.2. The Organ Distribution of the LDH Isoenzymes

Among the original observations of Markert and Moller were the facts that the LDH ISE patterns in the hearts of different mammals were different but more importantly that the patterns in various tissues of the *same mammal* were different and it is on this latter point that the clinical usefulness of LDH ISEs lies. As the data in Fig. 9.1a show the total LDH activity in the 5 organs and cell types are similar, however the ISE patterns are not. Some organs, eg brain and skeletal muscle, while not containing all the ISEs do show a broad spectrum of types. However others are much more narrow in composition, and of particular value diagnostically is the very high proportion of LDH 1 + 2 in the heart, and of LDH 5 in the liver.

These patterns develop fully only in late childhood but after this time they are of some diagnostic value.

	Liver	Heart	Skeletal Muscle	Brain	Red blood [Cells]
Total LDH Activity (U g^{-1} wet weight)	156	124	147	?	36
LDH Isoenzyme Distribution (% of Total)					
LDH1		>45		15–20	35–45
2		35–45	15–20	25–35	35–45
3	10–15	10–15	15–20	25–35	
4	10–15		15–20	15–20	
5	>45		40–55		

Fig. 9.1a. *LDH distribution among organs*

9.1.3. The Diagnostic Value of the LDH ISEs

∏ To be sure that you have understood this concept, answer the following question. Visualise a situation in a casualty department of a general hospital. An individual is brought in with one hand clutching his abdomen and the other his chest; he is in great pain and can tell you nothing.

(*a*) Explain why measurement of the total LDH activity in his serum will tell you little concerning the origin of his pain.

(*b*) With regard to the pattern of LDH ISE in his serum, what might you expect to find if the patient was suffering from a heart attack?

(*a*) Measurement of total LDH would not discriminate between a possible heart attack and a major liver complaint because in both cases a rise would be seen. This measurement is not a complete waste of time however since it will tell you if one of the major organ systems is generating the problem rather than say poisoning, a physical blow to his head or body, a perforated ulcer or any one of a number of other causes,

(*b*) With a heart attack LDH 1 + 2 should be raised and be the major contributors to the elevation of total LDH activity. We shall return to the use of enzymes in the investigation of the common heart attack later.

The LDH ISEs are a good example to study in a Unit of this type since they have proved to be of some value in investigations of quite a range of conditions. With regard to the heart, some conditions such as angina and pericarditis, both of which produce pain similar to a heart attack, can be distinguished from the latter by the lack of any sizeable change in LDH level. Other haematological problems, eg pernicious anaemia, cause an increase in all LDH types and are again therefore distinguishable from a heart attack.

LDH 1 tends to rise prior to transplant rejection and will rise in the cerebrospinal fluid as a consequence of central nervous system lesions such as thrombosis or haemorrhage.

LDH 4 + 5 tend to rise following liver conditions but only if the parenchymal cells (hepatocytes) are affected. Bile duct blockage produces no substantial change in LDH ISEs. Interestingly the location of the defect in hepatocytes can be confirmed by a study of another isoenzymic enzyme, since 2 out of the 3 alkaline phosphatase ISEs should show an elevation. Many malignancies result in a regression of certain aspects of cell structure and function to embryonic forms; this includes a tendency to anaerobic metabolism and the increased synthesis of the anaerobic M units. Some cancers are therefore revealed by an increase in LDH 4 + 5 activity.

These examples show quite clearly that the LDH pattern in serum. is of major value in improving organ specificity in the diagnosis of disease sites. A valuable point in this regard is that the serum ISE pattern is still organ specific even after the total activity has returned to normal.

∏   Which of the following conditions would result in increases in the serum activity of (*i*) LDH 1 and (*ii*) LDH 5?

(*a*) myocardial infarction (heart attack),
(*b*) angina,
(*c*) pericarditis,
(*d*) pernicious anaemia,
(*e*) kidney transplant rejection,
(*f*) brain haemorrhage,
(*g*) obstructive jaundice (bile duct blockage),
(*h*) liver cirrhosis (degeneration of liver cells),
(*i*) liver cancer.

(*i*) LDH 1 is significantly raised in (*a*), (*d*), (*e*), (*f*),

(*ii*) LDH 5 is raised in (*h*) and (*i*).

9.2. A SELECTED ORGAN – THE LIVER

9.2.1. The Basic Anatomy and Functions of the Liver

As a prelude to our discussion of this topic it will be useful to summarise the basic functions of the liver.

The liver is one of the largest of the body organs, averaging 1.4 kg in an adult. It receives blood from the gut via the hepatic portal vein and from the general circulation via the hepatic artery; and in line with this, its functions can be conveniently divided into those concerned with digestion and food processing, and those with a more general body role.

While the liver is structurally relatively simple it has a very wide range of essential metabolic functions of which the following are perhaps the most significant:

— the general interconversion of the products of the digestion of carbohydrates, lipids and proteins;

— the storage of glycogen, minerals and vitamins;

— the synthesis of many essential plasma proteins including the clotting factors and carrier proteins but not the antibodies;

— the metabolic breakdown or transformation of compounds (xenobiotics) regarded as foreign to the body; drugs, pollutants, pesticides etc;

— the destruction of old red blood cells and some bacteria by special (Kupffer) cells;

— the breakdown of haemoglobin from old red blood cells, and the excretion of the main product of this, bilirubin, into the gut as bile. While this breakdown occurs in the normal liver parenchyma cells (hepatocytes), transfer into the gut uses a system of bile capillaries and ducts.

∏   What in general terms, do you think this metabolic diversity might mean concerning the conditions that could develop in the liver?

It is a fact that a very wide range of conditions can develop in the liver. This is partly because it is an organ with an involvement in many essential aspects of metabolism, each of which can become defective, but also because it is an organ of such central importance that secondary effects frequently arise as a result of derangements occuring primarily elsewhere. A very useful feature of the liver is that its reserve capacity is considerable, it is quite possible to exist on only one third of a normal liver. The liver also has a remarkable capacity for regeneration by cell division and growth.

9.2.2. The More Common Diseases of the Liver

In order that our discussion of the role of enzymes in liver investigations is more meaningful it will be useful to outline the major types of condition that can affect the liver. Bear in mind that these conditions are not necessarily distinct and may be extremes of a spectrum of effects. Also important is the fact the the development of one condition (not necessarily in the liver) may induce the development of others (again either in the liver or not) by the production of toxins, general metabolic derangements etc.

(*a*)   Hepatitis (a general term for various inflammatory conditions; divisible into acute and chronic types). Acute hepatitis is usually viral in origin and the type B is particularly important because of its ease of transmission through body fluids and hence the possibility of outbreaks of infective hepatitis in hospitals and elsewhere, together with the overall seriousness of the clinical condition. Acute hepatitis may also occur as a result of infectious mononucleosis (glandular fever) or septicaemia (blood poisoning). By definition chronic hepatitis lasts at least 6 months and may arise following viral hepatitis or from alcohol abuse or drug reaction. Cellular destruction is limited in extent.

(*b*) Cirrhosis (while the term literally means 'tawny', and refers to the colour of the affected liver) more important features of cirrhosis are the destruction of large numbers of cells and the development of much fibrous tissue. To a certain extent the liver inflammation seen in hepatitis and cirrhosis are extremes of a continuous range, and cirrhosis may in fact develop out of hepatitis. Cirrhosis can therefore be induced by the same general type of agent as hepatitis, with differences in type, concentration, length of exposure or susceptibility of the individual determining the extent of the condition.

(*c*) Hepatic failure (a term indicating a metabolic failure of the liver parenchymal cells). This condition is important since patient survival rate in severe cases is only 20–30%. A wide range of causes exists, including viral hepatitis and xenobiotic reaction, with chlorinated hydrocarbons, benzene derivatives and therapeutic drugs such as paracetamol being of importance.

(*d*) Jaundice (a general term for diseases producing a yellow skin colour due to the accumulation of bilirubin and other bile products in the blood). Jaundice has three main origins:

— an excess of bilirubin produced due to a high rate of red blood cell breakdown (haemolytic jaundice);

— a derangement in liver cell metabolism leading to a failure to absorb bilirubin into the liver or excrete it into the bile (hepatocellular jaundice). Included within this group are certain inherited diseases in which defective metabolism exists;

— obstruction to the biliary system physically reduces or prevents bile flow; a condition called cholestasis (obstructive or post-hepatic jaundice).

(*e*) Tumours (cancers or 'growths'). These are unfortunately relatively common and may originate in the liver (primary) or result from the spreading (metastasis) of cancers from elsewhere (secondary). They can occur in the parenchymal or biliary systems of the liver.

∏   Let us take a very simple view of the various jaundice con-
    ditions for a moment and assume that they are distinct con-
    ditions, ie one will not induce another (which unfortunately
    is not the case in reality).

   (*i*)   Which of the 3 do you think is least likely to produce
           a release of liver parenchymal cell enzymes?

   (*ii*) Which of the 3 do you think is most likely to produce
           a release of enzyme from cells lining the bile canals?

   (*iii*) Do you think an inherited disease in which there is
           defective uptake of bilirubin into the liver or a defective
           excretion into the bile will result in the release of liver
           cell enzymes?

   (*i*)   The haemolytic jaundice arises because an excessive
           breakdown of red blood cells results in the produc-
           tion of more haemoglobin breakdown products than
           the liver and bile systems can cope with. The surplus
           accumulates in the blood and produces the jaundice.
           The liver cells themselves are normal and hence tend
           not to release enzymes into the circulation.

   (*ii*) Obstructive jaundice usually occurs because a gallstone
           or cancer develops in the bile capillaries or major
           ducts, and this leads to a regurgitation of bile into the
           blood system. Obviously the cells of the bile system
           are more likely to be damaged than those of the liver
           parenchyma and they could be identified as the site of
           trouble if enzymes specific for these cells can be mea-
           sured.

   (*iii*) The inherited diseases usually do not, in the first in-
           stance, produce generalised defects in the liver cells
           and therefore do not tend to cause the release of nor-
           mal liver cell enzymes.

9.2.3.   The Most Significant Diagnostic Enzymes

The liver has a very extensive range of metabolic roles, each of which could become defective. A great many enzymes exist which are specific to these pathways but most are either not available for assay when the condition develops, or the assay methods are unsuitable for routine use.

The majority of laboratories use relatively few enzymes in the course of the assessment of liver function, particularly ALP, AST, ALT, GGT and in some case LDH and 5'-nucleotidase. Disease specificity is improved by a consideration of some enzyme activity ratios but an important general principle that should be borne in mind here however, is that enzyme assays may be only part of a spectrum of investigations carried out during diagnosis. For the liver there is a particularly wide range of such other approaches as the standard textbooks of Clinical Chemistry (eg Varley *et al* (1980) will show.

The Individual Enzymes

(*a*) *Alkaline phosphatase*   This enzyme has a very wide distribution in the body with especially high activities in bone, placenta, kidney, gut mucosa and the bile system. Clinical assays on ALP have been performed for more than 50 years and organ specificity has been improved somewhat by the demonstration that bone, liver and placental isoenzymes can be distinguished by their stability to heat, urea and other conditions. Increases are found in hepatocellular disease and in cholestasis but are more substantial in the latter due to an increase in rate of synthesis as well as regurgitation into the blood as a result of duct blockage. In hepatitis the rise can be 2–3 times normal in the later stages of the disease, which is not particularly useful due to the wide distribution of ALP in other organs. In obstructive jaundice the rise can be more substantial especially in cholestatic types where it can reach 1400 U dm^{-3} (ie > 20 times the average normal). In mild cases the rise in ALP is a more sensitive indicator of cholestasis than the regurgitation of bile as shown by a measurement of bilirubin. This is particularly so if ISEs are studied since the bile pattern is different from that of the normal serum.

Increases of ALP also result from the so-called 'space occupying lesions' (carcinomata, abscesses, deposits of fibre, protein etc).

(b) *Aminotransferases* AST (GOT) and ALT (GPT)

These enzymes are present in every body tissue including the red blood cells. AST is especially high in heart and liver, with lower activities in skeletal muscle and kidney, ALT is in general of lower cell activity, but it is useful that its highest activity is reached in liver and kidney (Fig. 9.2a). Note that the differences in the data between this figure and Fig. 8.5a reflect the different methods used in the two investigations. The data in Fig. 8.5a are taken from Schmidt and Schmidt (1975), and those in Fig. 9.2a from Wroblewski (1958).

	Heart	Liver	Muscle	Kidney	Pancreas	Spleen	Lung
AST	1510	370	960	880	270	140	100
ALT	70	430	47	190	20	12	7

Fig. 9.2a.   *Organ distribution of aminotransferases*
$(U \times 10^3 \text{ g}^{-1}$ wet weight)

The aminotransferases show increases as a result of both cholestasis and hepatic cell damage (including cancer and alcohol abuse), and since moderate rises (1–8 times normal) can result from either state, the enzymes are of little diagnostic relevance here. However more dramatic rises (say 100 times) occur for AST in serious acute hepatitis; and this enzyme is obviously valuable in the diagnosis of this condition. In fact it is even more useful than this would imply since moderate changes occur very early in the development of the condition, certainly well before jaundice is seen. ALT is even better and more sensitive and the early diagnosis which becomes possible is of great value in investigating suspected outbreaks of infective hepatitis in hospitals and elsewhere.

Chronic hepatitis only produces moderate rises in aminotransferase activities but these rises show a useful relationship to the extent of the disease and recovery from it.

Significant use has been made of ratios of these enzymes. The distinction of liver and non-liver conditions is possible since increases of ALT (giving a moderate AST/ALT ratio) tend only to occur in the former. In the absence of significant ALT release (a high AST/ALT ratio) the AST probably originates from the heart.

With 40% of the AST being located in the mitochondrion and therefore requiring relatively substantial cellular damage before release, the extent of cell damage can be assessed from the AST/ALT ratio. Tissue recovery is assessable also with the ratio decreasing from > 2 in acute viral hepatitis as the individual recovers.

∏   The following table (Fig. 9.2b), shows the levels of some blood constituents in three clinical conditions (*i*), (*ii*), and (*iii*), which of the three sets of data do you think originate from mild viral hepatitis; acute hepatitis and recently developed obstructive jaundice?

[Remember that bilirubin is the main product of haemoglobin breakdown and is accepted, metabolised and then exported by the liver via the bile system. Note also that the blood contains large quantities of albumin which is synthesised by metabolically active liver parenchymal cells.]

	Reference [Values]	(*i*)	(*ii*)	(*iii*)
Bilirubin	<20 μmol dm^{-3}	85	17	140
Albumin	30–50 g dm^{-3}	31	41	44
ALP	30–120 U dm^{-3}	144	116	550
AST	<35 U dm^{-3}	1140	314	32

Fig. 9.2b.   *Blood enzyme levels in three clinical conditions*

Condition (*ii*) is mild viral hepatitis. In over half of the cases of mild viral hepatitis, liver cell damage is insufficient to impair substantially bilirubin metabolism and cell function. Therefore albumin synthesis and bilirubin levels tend to be within the normal reference limits. Mild cell damage causes the release of some AST but the increase is usually less than 10 fold. ALP is largely released from the bile system and is therefore within, but close to the top of, the normal limits.

Condition (*i*) is acute hepatitis. Cell necrosis leads to the liberation of considerable amounts of cell enzyme, with a peak activity of 10–100 times normal. Metabolic defects lead to a reduced cell metabolism as reflected in low albumen and raised bilirubin levels.

Condition (*iii*) is a recently developed obstructive jaundice. In obstructive jaundice significant release of bilirubin occurs due to the blockage. ALP levels rise to 3 times normal due to increased synthesis and regurgitation.

(*c*) γ *Glutamyl Transferase*

Relatively high concentrations of this enzyme are found in the kidney, pancreas, liver and prostate gland, which means that the specificity to the liver is not absolute but is nonetheless quite good. Of particular advantage is the fact that in contrast to ALP it does not increase in the serum following bone diseases as does ALP. The enzyme rises following both biliary obstruction and hepatocellular damage, with it being a more sensitive indicator of the former. The enzyme is a better indicator of cholestasis than ALP but a poorer indicator of hepatocellular changes than is the AST/ALT ratio. It has proved useful however in the initial diagnosis of cancers, the location of these in the liver being supported by a finding of increases in ALP and 5′-NT activities.

Undoubtedly its main use is in the confirmation of alcoholic abuse, it being a common characteristic of habitual alcoholics to deny their condition strenuously. The relationship between alcohol abuse and increased GGT is very strong, but it needs to be borne in mind that the rise in serum level is due to induction of increased synthesis and not just to increased liberation; other foreign compounds, eg barbiturate drugs, can act as inducers also and this can lead to false diagnoses.

∏   The myocardial infarction is a very important cause of heart attack and results in the liberation of enzymes from the damaged heart muscle. In about 50% of patients with an MI the serum GGT activity starts to increase after a few days. What does this suggest is happening as a consequence of the MI?

This finding highlights an important point in diagnosis, namely that one damaged organ can affect another. In this case the damaged heart is causing mild liver damage, resulting in the release of GGT. Apart from being of physiological and medical significance, this 'domino' or 'knock-on' effect can make diagnosis of the primary site of the disease much more difficult.

(d) *5'-Nucleotidase*   In most cases the changes in this enzyme parallel those of GGT and it could therefore be used as an alternative. The assay is however technically difficult and relatively insensitive and it is unlikely that the enzyme would be of any clinical significance were it not for the fact that it does not suffer from xenobiotic induction as does GGT.

Perhaps its main use is in the confirmation of the origin of ALP as liver rather than bone, bearing in mind that the ALP activity is elevated in bone diseases and in normal but growing children due to their high rate of bone metabolism. A high ALP/5NT ratio indicates a non-liver condition, a low ratio with high activities a liver condition. Liver and bone can be excluded with a finding of low ratio and low activities.

(e) *Ornithine carbamoyltransferase*   Apart from a low enzyme activity in the small intestine, the enzyme is virtually specific to the liver. It is the best enzyme for the identification of acute intermittent blockages of the biliary system.

∏ The following table (Fig. 9.2c), shows the serum activities for two clinical conditions (*i*) and (*ii*).

What is it about the data in column (*i*) that indicates that the condition does not originate in the liver?

In contrast, what is it about the data in (*ii*) that suggests the disease to be of liver origin?

Enzyme	Enzyme activities (U dm^{-3}) Reference Values	(*i*)	(*ii*)
AP	(5–14)	12	11
ALP	(10–170)	610	120
CK	(0–50)	15	30
AST	(5–20)	30	60
ALT	(5–25)	20	83
GGT	(6–30)	15	11
5′-NT	(0–50)	45	80

Fig. 9.2c. *Serum enzyme activities for two clinical conditions*

The raised ALP is the only really significant increase above normal. The normal 5′-NT and ALT suggests that it originates in another organ, eg the skeletal system.

Most significant are the high ALT (de Ritis ratio, AST/ALT, of 0.72) and the high 5′-NT activity; both enzymes being relatively specific to the liver. If the condition was a heart one the de Ritis ratio would probably be considerably above 1, and as we shall see later the CK value would be raised also.

SAQ 9.2a

(*i*)   List the major roles of the liver.

(*ii*) What is a xenobiotic?

(*iii*) What is the major chemical product of the degradation of haemoglobin?

(*iv*) Distinguish the three major types of jaundice.

(*v*)   5′-Nucleotidase is very specific to the liver; why is it not used more commonly in diagnosis?

(*vi*) State two distinct conditions that can result in an increase in LDH 4 + 5 activity.

(*vii*) Which enzyme shows a substantial rise as a result of persistent alcoholism? State one reason why correlating an increase in this enzyme with alcoholism needs to be done with care.

			(*i*)	(*ii*)

SAQ 9.2b

Two individuals with (*i*) haemolysis and (*ii*) an inherited metabolic defect of the liver called Gilbert's Disease, had the following serum enzyme activities. Normal values are given in brackets. Explain these results.

		(*i*)	(*ii*)
Bilirubin	(<20 μmol dm^{-3})	54	42
Albumin	(30–50 g dm^{-3})	40	46
ALP	(30–120 U dm^{-3})	106	30
ALT	(<35 U dm^{-3})	14	30

SAQ 9.2c

In cases of pre-hepatic (haemolytic) jaundice why is an increase in LDH commonly found? How could you confirm the probable origin of this enzyme?

SAQ 9.2c

SAQ 9.2d   The figures in column (*i*) of the following table were found in a typical case of obstructive
jaundice analysed soon after the condition developed. When the condition was left untreated
the values in column (*ii*) were obtained. Explain
the differences shown.

		(*i*)	(*ii*)
Bilirubin	(< 20 μmol dm^{-3})	150	200
Albumin	(30–50 g dm^{-3})	48	30
ALP	(30–120 U dm^{-3})	570	565
ALT	(< 35 U dm^{-3})	30	363

SAQ 9.2d

SAQ 9.2e

The table shows some typical enzyme activities for a case of acute viral hepatitis with column (i) obtained 1 day after the patient began to feel unwell, and column (ii) obtained 4 weeks after infection by which time he had apparently recovered.

		(i)	(ii)
Bilirubin	(<20 mmol dm^{-3})	30	18
Albumin	(30–50 g dm^{-3})	43	38
AP	(30–120 U dm^{-3})	118	110
ALT	(<35 U dm^{-3})	1140	32

What is the significance, with regard to ease of diagnosis, of the low bilirubin and high ALT activities in column (i)?

What is the significance of the figures in column (ii)?

SAQ 9.2e

9.3.   A SELECTED CONDITION – THE MYOCARDIAL INFARCTION

9.3.1.   The Nature of the Myocardial Infarction (MI)

The heart of a typical person is essentially a mass of muscle which is continually active throughout life, and when necessary is capable of very vigorous activity indeed. As such it is essential that it is kept well supplied with blood and consequently a number of serious problems can develop if the coronary circulation is faulty. A reduced oxygen supply will weaken, but not kill cells (a state known as ischaemia), and angina pectoris is a typical condition with this origin. More serious is the myocardial infarction or MI (often referred to as a heart attack or coronary) in which an area of tissue dies (an infarct). This is commonly due to a serious reduction in blood supply, usually as a result of blood vessel blockage. The condition is important because of its prevalence and seriousness. In the USA it is the cause of nearly 9% of all deaths between the ages of 25 and 34 and 25% of all those between 35 and 44; 140 000 deaths due to all heart conditions occur in the UK per year. Over half of all sudden deaths are due to an MI.

There are many other conditions affecting the heart, and they have a variety of origins and vary in seriousness. For the sake of brevity we will restrict ourselves to a study of the value of enzymes in the investigation of the MI.

9.3.2. The Investigation of the Myocardial Infarction

Reliable diagnosis of the presence and the extent of an MI is of paramount importance because in serious cases careful treatment in a coronary care unit should at least double the chances of survival. Diagnosis of a less serious MI gives time for the introduction of medical treatment and adjustments in personal life-style, which can substantially extend life expectancy. Important in this context is the fact that at least 20% of infarcts do not produce pain, and in going unnoticed do not give the advance warning a mild MI usually gives. The incidence of painless MI rises with age which is unfortunate since the tendency to develop an MI rises with age also. Another crucial point is that while the major physical manifestation of an MI is pain, it may take many forms and locations, and distinguishing an MI from say shoulder joint rheumatism, pulmonary embolism, abdominal pain etc is necessary but difficult. Roberts *et al* (1975) gives a useful review of the main approaches to the investigation of MIs.

While radioisotope-based whole-body imaging techniques are rapidly developing, at present two major approaches to the investigation of MIs are commonly used, one being the electrocardiograph (ECG) study, which relies upon changes in the electrical signals picked up from the contracting heart. The other is a study of the changes seen in the blood as a consequence of the release of enzymes from the dying cells. While the ECG investigation gives the most rapid return of information, the enzyme assays are nonetheless of great value since cardiac massage given to restore life, previous attacks and certain cardiac drugs (eg digitalis) can significantly distort the ECG pattern. Unfortunately a reasonable number of patients developing an MI already have a known heart problem and may be on such drug therapy. Enzyme measurements have additional merits in particular that by careful choice of investigation they can:

— confirm the presence of an MI and distinguish it from other conditions (not just of the heart), which might be causing the pain,

— show the presence of multiple, sequential MIs,

— show the time elapsed since the MI,

— indicate the chances of survival.

∏   How can we distinguish between the conditions of angina
         pectoris and myocardial infarction? Which do you think is
         likely to produce the more significant enzyme changes?

         Both these conditions result from a reduction in blood sup-
         ply to the cells but whereas in an MI the cells actually die,
         in angina they merely change physiologically. In the case
         of angina comparatively little enzyme release occurs and in
         many cases the levels of those enzymes usually measured
         may not exceed the upper limit of normal.

The pattern of serum enzyme changes following a typical MI are
shown in Figs. 9.3a and b, from such a complex series of changes a
great deal of information can be obtained. The production of a full
set of profiles of this type requires a considerable amount of effort
and it is not practicable for most laboratories to carry out such a
study on each patient. Laboratories tend to be selective in the assays
they undertake and in fact considerable variation is seen between
laboratories in the enzymes chosen for measurement, the number
of measurements made and the methods used for the assays.

Enzyme	Adult upper limits of normal (units dm^{-3})	Normal increases (fold)	Normal times to maximum rise (h)	Normal times to return to normal (days)
Total CK	100	5–8	20–30	2–3
CKMB	6	5–15	24	2–3
AST	25	3–5	24–48	4–6
Total LDH	290	2–4	48–72	7–12
HBDH(LDH1)	100	2–4	48–72	6–14

Fig. 9.3a.   *Serum enzyme changes following myocardial
infarction*

Π The chances of a patient surviving an MI improve substantially if he survives the first few days after an attack; and because of this it is useful to know what time has elapsed since his attack. Why do you think a single measurement is of limited value in this context?

The problem with a single measurement is that while a raised value may indicate the presence of an MI it cannot show a trend, and there is no way of knowing whether it lies on the rising or falling part of the curve. An inspection of Fig. 9.3b will show that a AST value of twice normal could occur either 8 h or 2.5 days after the infarction. Thus ideally several sequential measurements ought to be made.

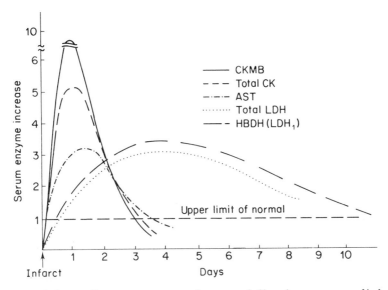

Fig. 9.3b. *Serum enzyme changes following myocardial infarction*

An important point influencing the choice of enzymes to be measured in these patients is that it is unfortunately the case that none of the enzymes shown in Fig. 9.3a are specific to the heart. Most are present in other organs and can be released into the circulation as a result of a wide range of conditions. However we shall see that there are ways to reduce this problem, but nonetheless it does mean that multiple investigations are necessary.

While other enzymes do show changes in response to an MI, the four enzymes shown and discussed below have proved particularly useful. Increases in serum level of these enzymes occur in $> 90\%$ of patients with an MI and a good correlation with prognosis is known for all except LDH. In general a 10 fold rise in activity above the normal level would suggest a 50% chance of mortality.

9.3.3. The Value of Individual Enzymes in the Assessment of the Myocardial Infarction

Let us now look at the individual enzymes and assess their particular merits.

At first sight CK would appear to be the most useful enzyme since it begins to rise very soon after the MI and on average produces an easily measured 5-fold rise in activity. The rise can in fact be as high as 20 times the upper limit of normal and the activity frequently peaks within 24 h after the infarct. However the heart does not have the highest concentration of CK among the large body organs with the ratio of activities for skeletal muscle: heart: brain: gut being $30:7:2:1.5$. Thus effects on other organs, particularly striated muscle can cause significant rises in CK activity. In the case of straited muscle common activities such as strenuous exercise, fevers, intramuscular injections or blood sample removal, as well as surgery and muscle-wasting diseases such as the dystrophies, cause the release of CK. It is ironic that the removal of a sample for enzyme assay could in itself alter the CK level and make subsequent measurements more difficult to interpret. Prolonged shock is another confusing state with a rapid rise in CK level to over 20 times normal being possible.

A final factor weighing against the use of CK is that it is relatively unstable; and indeed the problems this generates are so significant that it was selected as an example in our previous discussion in (5.1.4). The assay procedure is complex, error prone and expensive due to the nature and number of reagents involved (see 3.2.2).

However despite these problems laboratories are still interested in CK because it is an enzyme with isoenzymic forms which are of considerable value. The CK molecule is a dimer of M and B units (for muscle and brain). The CK MB form is almost totally restricted to the heart (Fig. 9.3c) which gives a considerable improvement in organ specificity as illustrated by Fig. 9.3d which shows the normal values for serum CK and CK MB, and the changes following a number of diseases. The assay itself is relatively straightforward since the 3 forms can be separated electrophoretically and measured by a fluorimetric reaction. Alternatively, commercial kits are available which are based upon the blocking of the M sub-unit with antibodies, which will remove all of the activity due to CK MM and reduce the MB activity to half. Residual activity should be due to the remaining MB activity since the BB form is not found in the serum unless the blood brain barrier is damaged. Unfortunately the method is said to lack precision and sensitivity, and this latter point is particularly important since only 5-10% of the serum CK activity is due to the MB form, the MM being overwhelmingly predominant.

Organ	Total CK	CK MM	CK MB	CK BB
Skeletal Muscle	3000	3000	3	0
Heart	700	600	100	0
Liver	3	0	0	3
Brain	200	0	0	200
Gut	150	0	0	150
Lung	15	10	0	5
Kidney	10	1	0	9
Uterus	10	2	2	6

Fig. 9.3c. *CK distribution among organs*
Values are in U g^{-1} wet weight

Disease	Organ affected	Total CK $(U\ dm^{-3})$	CKMB $(U\ dm^{-3})$
Acute pancreatitis	Pancreas	113	0
Stomach cancer	Stomach	283	5
Cerebral haemorrhage	Brain	92	0
Central nervous system tumours	Brain/Spine	226	8
Myasthenia gravis (nervous degeneration)	Limb muscle	146	0
Polymyosotis (inflammation)	Limb muscle	760	23
Myocardial infarction	Heart	500	15–100
Normal Serum Activities		50	10

Fig. 9.3d.   *The effect of selected diseases on serum total CK and CK MB*

∏   An assay method for CK MB has recently been marketed in which the serum is treated with an antibody to the M unit. Since this antibody inhibits the enzyme activity, (*a*) what will the residual activity be due to?

In the second stage of the assay an anti-antibody is added which precipitates all molecules carrying the anti-M antibody, (*b*) Explain what the residual activity will be due to and why is this stage useful? (*c*) Show how the CK MB activity is calculated from the information obtained from these two treatments.

(*a*) CK BB and CK MB since the M unit has been inhibited. The activity of the CK MB ISE is due to the B unit that it contains. (*b*) Residual activity is due to CK BB, since the CK MB will have precipitated out by the anti-antibody. This is useful since although the normal levels of BB in blood are zero, patients are by definition frequently abnormal and sufficient BB may leak through a damaged blood brain barrier to interfere with other assays. (*c*) CK MB activity = (activity from stage 1 minus activity from stage 2) × 2 since stage 1 measures BB + half MB, and stage 2 removes BB activity leaving a residual activity due to half of the MB molecule.

Perhaps because the results of measurements of the MB form of CK are not clouded by super-imposed changes in other forms, measurements of the MB activity are more sensitive to changes in cardiac physiology. Of particular advantage are the facts that CK MB level can show a rise in activity as early as 6 h after the infarct and can peak up to 12 h earlier than total CK activity. It will also show the presence of a series of infarctions, and being restricted to the heart is not influenced by intra-muscular injections or blood sampling procedures (Fig. 9.3e).

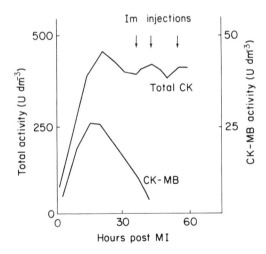

Fig. 9.3e. *Response of CK measurements to intra-muscular injections*

Unfortunately there are some non-MI processes which directly affect the heart and thereby cause a rise in CK MB activity. External cardiac massage, heart surgery and the investigative technique known as coronary angiography in which radio-opaque dyes are injected into the coronary vessels in order to assess their condition are among these.

Prior to the measurement of CK MB, attempts were made to improve organ specificity for the heart by obtaining the CK/AST ratio. Ratios of 2–9 were indicative of a heart origin whereas the higher activity in the muscle would give ratios of 9–50 if this were the source. The need for this determination is now much reduced.

Perhaps the enzyme most widely used for MI detection in hospital laboratories is AST, since the increase begins reasonably soon (6–8 h) after the infarct, can last for a useful 6–8 days or so, and the assay procedure is relatively straightforward. The increase is commonly up to 5 fold but can be as high as 25 fold. If the timing of the infarct is known (eg from the onset of the pain) then there is a good correlation between the size of the increase in AST and the size of the infarct.

The other main aminotransferase (ALT) is not present in the heart and its main value is a negative one; a rise in ALT indicating a liver rather than a heart condition.

The other enzyme that is studied with any frequency is LDH, in particular the hydroxybutyrate dehydrogenase (HBDH) activity, characteristic of LDH1. This latter assay has the advantages of being more sensitive than that of total LDH, showing a considerable improvement in organ specificity and prolonging the period of demonstrable enzyme changes for up to 15–20 days from the time of the MI. While this is said to be quite useful, since the early rising enzymes (CK and AST) tend to return to normal within 2–6 days, in practice the assay is suspect, the additional information not often required and the assay therefore relatively uncommon. An expression sometimes used is the LDH/HBDH ratio which is 1.2–1.6 in a normal serum and less than 1.2 following an MI.

SAQ 9.3a

A measurement of total CK activity was undertaken 1.5 days after a probable MI.

You would like to confirm the time of onset of the MI – what other measurement could you carry out?

The patient is very ill and you would like to monitor the possibility of repeat infarcts over the next few days. How could you do it?

SAQ 9.3b

There has been a time lapse of 10 h since the onset of chest pain. Out of the following list which pair of enzymes would be most appropriate to determine whether the pain was due to an MI?

(*i*) CK and LDH1;

(*ii*) CK and AST;

(*iii*) CK and ALT;

(*iv*) AST and ALT;

(*v*) AST and LDH1.

SAQ 9.3b

SAQ 9.3c

There has been a time lapse of about 3 days since the onset of chest pain. Which enzyme(s) from the list below would you expect to see raised if (*i*) the pain was due to an MI, and (*ii*) it was due to angina.

ALT,
AST,
CK MB,
Total CK,
LDH1.

Summary

In this Part you will have studied some specific examples of the value of enzyme assays in diagnosis. Three different viewpoints have been taken, namely;

— a particular enzyme and its diagnostic value;

— a particular organ and its defects;

— a particular condition and the methods by which it is investigated.

Objectives

You should now be able to:

● list the structures of the isoenzymic forms of LDH, describe the distribution of these isoenzymes among selected organs, discuss the value of this distribution in the investigation of diseases of these organs, and list various disease states producing changes in serum isoenzyme patterns;

● list 5 distinct functions of the liver, define hepatitis and cirrhosis, distinguish the 3 main types of jaundice, and discuss the value and limitations of selected enzymes in the investigation of liver diseases;

● describe the nature of an MI, justify the importance of detailed investigations and monitoring of it, and discuss the value and limitations of selected enzymes in the investigation of the MI.

Self Assessment
Questions and Responses

SAQ 1.2a

Which of the following statements is correct? The activation energy of an enzyme reaction is:

(*i*)   the energy needed to activate the enzyme for reaction;

(*ii*) the energy needed to activate the substrate for reaction and form the transition or intermediate states;

(*iii*) the energy liberated during the reaction;

(*iv*) the energy retained by the products;

(*v*)   the energy provided by the products.

Response

Statement (*ii*) is the answer here. The chemistry of reactions is generally such that complex intermediates or transition states involving reactive groups on the enzyme surface are present during the conversion of substrates to products. Even if the substrates have a higher energy level than the products these intermediate states will have higher energy levels than those of either the substrates or products. A so-called 'activation energy' must therefore be acquired by the substrates before formation of products is possible. None of the other statements is even partly correct.

****

SAQ 1.2b
The following is a diagram representing the energy changes occurring during the reaction.

$$H_2 + 0.5\,O_2 \rightarrow H_2O$$

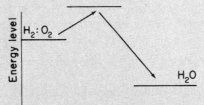

Draw in the curve representing the changes occurring in the presence of platinum as a catalyst.

Response

As a catalyst you would expect platinum to reduce the activation
energy requirement to give a profile such as the one below.

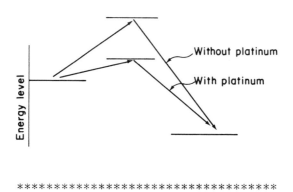

****

SAQ 1.2c	The following table shows the activation energy for the hydrolysis of fructose-containing poly-mers:

sucrose + H_2O → glucose + fructose

raffinose + $2H_2O$ → glucose + 2 fructose

Reaction (hydrolysis of)	Catalyst	Activation energy kJ mol^{-1}
(*i*) sucrose	H^+	108.7
(*ii*) sucrose	malt sucrase	54.4
(*iii*) sucrose	yeast sucrase	46.0
(*iv*) raffinose	yeast sucrase	46.0

→

SAQ 1.2c
(cont.)

1. Which of these reactions (*i*) and (*ii*) is likely to proceed more readily?

2. Note the difference in energy value between (*ii*) and (*iii*) and the similarity of value for (*iii*) and (*iv*). Are these data suggestive of the enzyme, or the substrate and the particular reaction being the major determinant of the activation energy value?

Response

1. In the absence of any other influencing factor, reaction (*ii*) should proceed more readily due to its significantly lower activation energy requirement. At any given time and temperature, more substrate molecules will achieve the 54.4 kJ mol^{-1} requirement than the 108.7 one.

2. These data, together with many other experiments, suggest that the activation energy value is largely determined by the enzyme.

****

SAQ 1.3a

The enzyme urease catalyses the hydrolysis of urea to carbon dioxide and ammonia.

$$(H_2N)_2CO + H_2O \rightarrow 2\,NH_3 + CO_2$$

The compounds thiourea $(H_2N)_2CS$, hydrazine $(H_2N)_2$ and semicarbazide $H_2N.NH.CO.NH_2$ all give an inhibition in excess of 95% whereas acetamide $H_2N.CO.CH_3$ barely inhibits at all.

$\longrightarrow$

SAQ 1.3a **(cont.)**	(*i*) Examine the structures of these molecules carefully and explain why acetamide is a non-inhibitor.
	(*ii*) What type of inhibition are these compounds exhibiting?
	(*iii*) Why are studies on the structure of inhibitors useful in investigations of enzyme reaction mechanisms?

Response

The formula of urea can be written so:

$$H_2N \diagdown_{\diagdown} C = O$$
$$H_2N \diagup$$

and it would seem likely that the enzyme and substrate join together via the two amine groups since this would allow thiourea, hydrazine and semicarbazide to act as competitive inhibitors.

(*i*) Acetamide has one of the amine groups replaced by a methyl and hence is unlikely to be able to bind to the active sites of the enzyme because of an unsuitable size, shape or charge distribution on the methyl group.

(*ii*) Thiourea, hydrazine and semicarbazide compete with the normal substrate for attachment to the enzyme active sites and are thus said to be competitive inhibitors.

(*iii*) Studies of the structures of compounds which are capable of competitive inhibition and those which are not, are of great value in understanding the substrate/protein interaction at the active site(s) of enzymes.

****

| SAQ 1.3b | Select from the following list those feature(s) that apply to the prosthetic groups involved in some enzyme reactions. |

Prosthetic groups

(*i*) can be distinguished from coenzymes by their solubility in the cell cytoplasm and other aqueous systems;

(*ii*) are normally proteins;

(*iii*) are present at the active site and covalently attached there;

(*iv*) are intimately involved in the enzyme reaction;

(*v*) carry groups from one reaction to another (ie from one enzyme to another);

(*vi*) are inorganic ions.

Response

Statements (*iii*) and (*iv*) are the correct answers. A good example of a prosthetic group is haem, which is covalently attached to the active site of enzymes such as catalase where its iron atom functions in the enzyme reaction. This attachment means they are insoluble. (*i*), (*vi*) and of course (*ii*) are therefore incorrect. Again being attached to enzymes they cannot function as carriers in the sense expressed in (*v*). This function is performed by the soluble coenzyme molecules.

****

SAQ 1.3c

Enzymes such as alkaline phosphatase can remove phosphate from a wide range of substrates
such as those illustrated schematically below.
P represents phosphate and the other letters a
range of possible chemical structures.

A—P   A—B—P   A
                         \
                          B — P
                         /
                        C

— What is the name given to this type of specificity?

— Why might the enzyme be unable to react
with a substrate of the following form?

A   B
 \   /
  \   /
   \   /
    P
   /   \
  /   \
 D   C

Response

An enzyme capable of metabolising a variety of related compounds
which all have a chemical group in common is said to show group
specificity. In this case the alkaline phosphatase is specific for the
phosphate group.

A group specific enzyme may not necessarily be able to metabolise
all compounds containing a particular group and one reason for
a failure to react could be the inaccessibility of the bond to be at-
tacked. In the hypothetical example shown the phosphate is shielded
by groups A, B, C and D, and the large size of the enzyme molecule
might prevent access to the phosphate group.

****

SAQ 1.3d	If a compound which has stereo-isomeric forms is synthesised chemically then both isomers tend to be produced giving a so-called 'racemic' mixture. How might enzymes be used to isolate just one of these forms?

Response

The racemic mixture could be treated with a stereospecific enzyme to convert one of the isomers to another compound which might be relatively easily separated from the remaining isomers eg:

$$\text{DL-amino acid} \xrightarrow{\text{L-amino oxidase}} \text{D-amino acid} + \text{another product}$$

A problem with this approach is the relative shortage of suitable enzymes. With natural compounds one isomer tends to be present in considerable excess, more than 95% of amino acids found in organisms are L-forms for example, and hence while this is the isomer that it is likely you are trying to synthesise for experimental purposes, most enzymes have evolved to *use* this isomer. It might be quite difficult and expensive to obtain the enzyme for the alternative isomer.

An answer to this problem might be to synthesise the required isomer by using a suitable enzyme instead of chemically. It is probable that the enzyme active site will have such a rigorous shape that the reaction will only occur in one orientation and thus only one isomer will be produced.

****

SAQ 1.3e

> What might the benefit be to a pathogenic micro-organism in having a relatively high proportion of uncommon stereo-isomers of amino acids in its cell wall? Consider the ways in which the host might try and destroy this pathogenic organism.

Response

As might be expected animals have evolved several quite complex mechanisms for counteracting the invasion and proliferation of pathogenic viruses, bacteria, fungi etc. While we need not go into details of these mechanisms, it is the case that enzymic attack on the cell wall of the pathogen is commonly involved.

Since the vast majority of the host's enzymes will have evolved to metabolise the common isomeric forms of stereo-isomeric compounds (ie the L-amino acids and D-sugars), having the uncommon form in the cell wall might reduce the extent of enzymic attack.

****

SAQ 1.3f

> This SAQ is perhaps a rather more difficult problem but think carefully about it because it does highlight the far-reaching consequences of the stereospecificity shown by enzymes.
>
> The following equation shows a simplified version of an important reaction occurring in aerobic respiration, that is the conversion of citric acid to 2-oxoglutaric acid. $\longrightarrow$

**SAQ 1.3f
(cont.)**

$$
\begin{array}{c}
\text{COOH} \\
|\\
\text{CH}_2 \\
|\\
\longleftarrow\text{----HO-C-COOH----}\longrightarrow \\
|\\
\text{CH}_2 \\
|\\
\text{COOH}
\end{array}
\qquad\longrightarrow\qquad
\begin{array}{c}
\text{COOH} \\
|\\
\text{CH}_2 \\
|\\
\text{CH}_2 \\
|\\
\text{C}=\text{O} \\
|\\
\text{COOH}
\end{array}
\; + \; CO_2 \; + \; 2H
$$

Citric acid   2-Oxoglutaric acid

If you look at the formula of citric acid it would appear that the molecule is symmetrical on either side of the dashed line drawn across it. However subtle radioisotope labelling experiments have shown that the enzyme actually metabolises only one particular end of the molecule. Why do you think this so-called 'Ogston Effect' occurs? Bear in mind that it is thought that the enzyme attaches to three groups in the citric acid molecule and you might like to sketch out the possible orientations of the citric acid molecule on an enzyme surface represented like this.

Enzyme surface

Attachment site

Response

If the enzyme does in fact attach to three groups in the citric acid molecule, then there is only one orientation for that molecule on the enzyme surface. You might like to check this with sketches or even with simple models to confirm it for yourself. It means in practice that only one particular group in the citric acid will be metabolised and the enzyme does *not* therefore treat the molecule as being symmetrical.

To simplify the discussion we can assign letters to the various groups of the citric acid as shown below, and it is then perhaps more apparent that only one face of the citric acid pyramid has the required arrangement of groups.

Citric acid	Representation of citric acid	Required arrangement of groups on the active site

****

SAQ 2.3a	You are investigating the activity of an enzyme in a blood sample by using a spectroscopic technique to follow the appearance of reaction product. You have produced the data supplied in Fig. 2.3c (i) for a calibration curve to relate absorbance to product concentration.

In your actual experiment you are able to follow the accumulation of product directly by the absorbance changes produced over a period of time when 100 μl (0.1 cm^3) of sample are used, Fig. 2.3c (ii)

Calculate the enzyme activity cm^{-3} of blood sample.

(i)		(ii)	
Absorbance of reaction product		Changes in absorbance during the reaction	
Concentration (μmol cm^{-3})	Absorbance	Time (s)	Absorbance
10	0.065	2	0.070
30	0.195	4	0.135
50	0.325	6	0.200
70	0.455	8	0.265
90	0.585	10	0.320
110	0.715	12	0.375
		14	0.420
		16	0.460
		18	0.500
		20	0.530
		22	0.560
		24	0.590
		26	0.615
		28	0.630
		30	0.645

Fig. 2.3c

Response

When plotted, the data of Fig. 2.3c (i) should be similar to Fig. 2.3d (i) and from the calibration graph a slope can be determined as shown and a value close to 0.0065 cm^3 μmol^{-1} should be obtained.

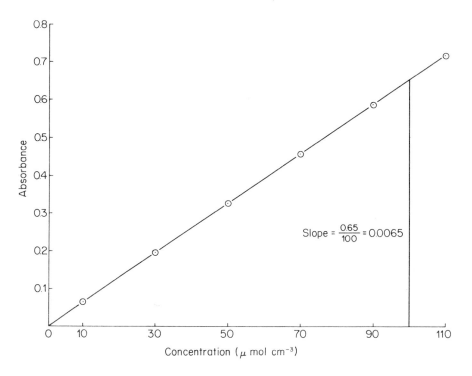

Fig. 2.3d(i).   *Plot of data from Fig. 2.3c(i)*

As explained in the text it is usual to determine enzyme reaction rates as initial values which are obtained at the beginning of the reaction before various factors begin to adversely affect the system and produce a decline in rate.

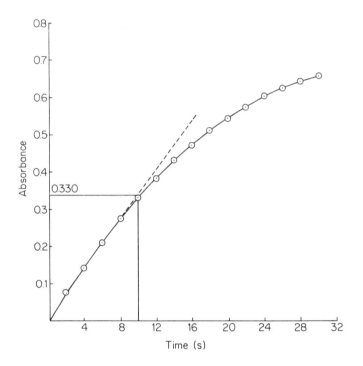

Fig. 2.3d(ii). *Plot of data from Fig 2.3c(ii)*

An inspection of Fig. 2.3d(*ii*) should show a linear reaction rate occurring for nearly ten seconds and hence it would be possible to measure a slope from this directly. However in practice initial linear rates are sometimes shorter than this and in any case true experimental data rarely fit on graph lines as well as these do. Consequently it is good practice to draw the best straight line through the data and determine the initial velocity from this. Such a line has been drawn and an absorbance change of close to 0.330 during the first ten seconds has been obtained. Using this data and Eq. 2.3e the following calculation can be done:

$$\text{Enzyme activity} = \frac{\Delta A}{\Delta t} \times 60 \times \frac{1}{k} \times 10^6 \ \mu\text{mol cm}^{-3} \text{ min}^{-1} \quad (2.3\text{e})$$

$$= \frac{0.330}{10} \times 60 \times \frac{1}{6.5 \times 10^{-3}} \times 10^6$$

$= 0.305 \times 10^3 \ \mu\text{mol cm}^{-3} \text{min}^{-1}$ from 0.1 cm^3 blood

$= 3050$ International Units cm^{-3} of blood sample (assuming 1 molecule of product derives from 1 molecule of substrate)

****

SAQ 2.5a   The following is a graph illustrating the rate of respiration by a yeast culture.

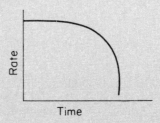

Give three of the possible explanations for the decline in rate with time.

Response

In reality a wide range of explanations are possible but perhaps the three most obvious are:

(i)   the substrate is being consumed and enzyme activity and hence respiration rate will fall when the enzymes are no longer saturated with substrates;

(ii)   the product is inhibiting the reaction by affecting the reaction equilibrium, or by more direct inhibitory effects on the enzymes or other components of the system. Under anaerobic conditions alcohols are produced by respiring yeast which readily denature enzymes and this is one factor that limits the amount of alcohol produced by industrial fermentation processes;

(*iii*) the enzymes and/or other cell structures are degrading due to their inherent instability leading to a reduction in cell metabolic activity or even to cell death.

****

SAQ 3.3a

For each of the following give:

(*i*) a reason why direct measurement of the product of an enzyme reaction might not be possible;

(*ii*) a potential disadvantage of the measurement of enzyme reaction product by chemical derivatization;

(*iii*) an advantage of the enzyme derivatization technique compared with chemical derivatization techniques;

(*iv*) a disadvantage of those enzyme derivatisation techniques that involve more than one derivatising enzyme;

(*v*) an advantage of luminometry as a measuring technique for derivatives of reaction products;

(*vi*) a disadvantage of using artificial substrates in measurements of enzyme reactions.

Response

(*i*) The product might not absorb visible light or ultra-violet radiation, or indeed may be more amenable to measurement by another technique.

(*ii*) In many cases the derivatizing reagent will inhibit or otherwise affect the enzyme itself so that if it is added to the reaction system only a single measurement will be obtained. The introduction of this step will increase the error involved in the measurement.

Many chemical derivatizing reactions lack specificity.

(*iii*) In most cases the enzymes added as derivatizing reagents are not inhibitory to the enzyme being measured in the sample and hence a continuous response will be obtained. This will allow a much more accurate determination of reaction rate to be made.

The continuous removal of reaction product will prolong the period before the reaction reaches equilibrium, again allowing a more accurate determination of rate to be made.

(*iv*) The introduction of a complex chain of derivatizing reactions markedly increases the error in the assay, partly by an increase in the number of opportunities for the normal sources of laboratory error but also by a tendency for such systems to develop a lag phase. The presence of a lag phase makes determination of a true initial reaction rate much more difficult.

The cost of the assay can rise dramatically since each enzyme in the chain must have an activity at least 10 times that of its predecessor in order to avoid rate-limitation effects.

(*v*) Some luminometric assay systems have very low detection limits (down to 10^{-13} mol dm^{-3} for ATP), and frequently have linear responses extending over several orders of magnitude of concentration.

A major advantage of luminometry is that is that it provides a completely different experimental approach to enzyme assay which might be suitable for a reaction which is not amenable to either direct measurement, or to the generation of other end products by chemical or enzymic derivatization.

(*vi*) By definition an artificial substrate is not the naturally occurring one but is chosen because its disappearance, or the appearance of the reaction product, can be easily measured. However the enzyme is unlikely to react quantitatively, and perhaps not even qualitatively in the same way as it would with its natural substrate. While this may not be important if comparative studies using the same reagents and conditions in all experiments are made, it can make inter-laboratory comparisons more difficult.

****

SAQ 4.2a

The list below contains a series of definitions of some measuring techniques which can be used to follow changes produced by the consumption of substrate or generation of product during the course of an enzyme reaction. The names of the techniques are also given in a second list, and you should match each definition with it's appropriate name.

The Definitions

(*i*) The generation of an emf in a cell containing indicating and reference electrodes, by ions in the solution.

(*ii*) The collection of gases produced by the reaction and the measurement of gas volume.

$\longrightarrow$

**SAQ 4.2a
(cont.)**

(*iii*) The change in resistance of a solution during an enzyme reaction.

(*iv*) The measurement of the change in optical rotation of a solution during the course of an enzymic reaction.

(*v*) The measurement of the temperature changes occuring during the course of a reaction.

The Names

manometry
calorimetry
polarography
polarimetry
potentiometric measurement
conductance measurement.

Response

(*i*)   potentiometric (or emf) measurement
(*ii*) manometry
(*iii*) conductance measurement
(*iv*) polarimetry
(*v*)   calorimetry

Note that polarography is an important technique included in the list of names but not given in the definitions which form the basis of the question. It involves the diffusion of materials present in a solution across a permeable membrane and their oxidation or reduction at an electrode surface. In biochemical situations it is most commonly used to measure dissolved gases.

****

SAQ 4.2b | State one problem associated with the measurement of enzyme reactions using each of the following techniques:

(*i*) microcalorimetry;
(*ii*) manometry;
(*iii*) conductance measurements.

Response

Each of these techniques has a number of problems which have prevented their widespread use, some of these are to do with the technique itself, and would be eliminated once the system was established and calibrated. Others are involved with applications of the techniques to biological specimens and the requirements of speed, minimal cost per assay, and minimal technical and specialist skill, which are of great importance in hospital laboratories.

(*i*) While the technical problems involved in measuring the small temperature changes occurring during the reaction have been overcome, and the instrumentation can be automated, current systems require rather long measurement times of 15–20 minutes per sample. The technique could however develop into one of considerable value since it should be applicable to virtually every reaction, which will increase the usefulness of any specialist skill required, and increase the cost-effectiveness of the instrumentation. It is also suitable for coloured and turbid solutions which is of importance in hospital laboratories; and another merit of lesser significance is its non-destructive nature allowing sample recovery.

(*ii*) This technique suffers from the requirement to have reasonably large gas volume changes before an accurate measurement can be made. This necessitates either a large volume of enzyme sample or a long reaction time, neither of which is compatible with clinical work. It also lacks widespread applicability compared for example, with spectroscopic assays.

(*iii*) The main problem with conductance measuring techniques is the determination of the small changes occurring in the presence of the high background conductance due to the ionic species already in the sample, the buffers and the reagents involved in the assay. The levels of these ionic species will differ among biological samples resulting in a variation in this background level, and an increase in the complexity and error of the assay. This technique also lacks widespread applicability.

****

SAQ 5.1a

> (*i*) State *four* examples or situations in which biological specimens including routine blood specimens, might only be available in small quantities.
>
> (*ii*) Very briefly describe *two* ways in which we can increase our ability to measure the enzymes in these small samples.

Response

(*i*) While the adult human body contains about a gallon (5 dm^3) of blood, there are a number of situations in which much smaller quantities are present, for example in fetuses and young babies. Alternatively access to the main blood vessels might be difficult as in severely burnt, grossly obese or geriatric patients. In such individuals small volumes of capillary blood might be the only material available.

The other main type of specimen available in small amounts is the tissue biopsy, it is perhaps self-evident that when taking pieces of living tissue under anaesthetic the surgeon will remove the smallest sample compatible with laboratory requirements.

(*ii*) In all measurements of enzyme activity we are relying on the catalytic nature of enzyme activity to generate sufficient product to enable us to measure the very small number of enzyme molecules present in the sample.

If normal methods are still not sensitive enough it might be possible to increase the sensitivity by using alternative substrates or more sensitive techniques such as fluorimetry, luminometry or radio-isotope labelling. Alternatively enzyme cycling (see Fig. 3.2k) might be employed.

With regard to tissue biopsies it is common to increase the quantity of available enzyme by increasing the cell number using tissue and cell culture techniques in conjunction with the use of the very sensitive assay methods just mentioned.

****

SAQ 5.1b	Adenylate kinase, released from cells which become damaged during or following the taking of blood samples, interferes with creatine kinase assay, (Fig. 5.1a). How do you think the release of another type of enzyme, the virtually omnipresent ATPases, would affect this assay? They carry out the following reaction:

$$ATP + H_2O \rightarrow ADP + PO_4^{3-} + H^+$$

Try and answer this question from memory before referring to Fig. 3.2j, in which the CK assay system is illustrated.

Response

The effect of this enzyme is to reduce the level of ATP in a solution by hydrolytic conversion to ADP and inorganic phosphate. Since both the derivatization system illustrated in Fig. 3.2j and the newer luminometric approach, act by measurement of the ATP generated by CK, the presence of an ATPase will produce an apparent reduction in CK enzyme activity.

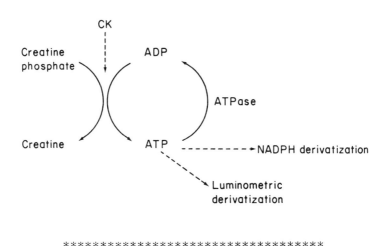

****

SAQ 5.1c   Briefly describe *two* of the possible effects on enzyme assays of the use of anticoagulants.

Response

The following are perhaps the most important of such effects. It should be noted that fortunately none of the common anticoagulants produces all these effects, and furthermore few enzymes are susceptible to all these effects.

(*i*) A movement of ions between the extracellular fluid and the blood which results in osmotic water movements and hence changes in the concentration of enzyme molecules in the blood.

(*ii*) Direct enzyme inhibition.

(*iii*) Indirect enzyme inhibition by removal of required cofactors.

(*iv*) Apparent enzyme inhibition due to interference with the assay system.

****

SAQ 6.1a   Fig. 6.1c shows typical reaction profiles for:

(1) a turbid sample;
(2) a normal, clear sample.

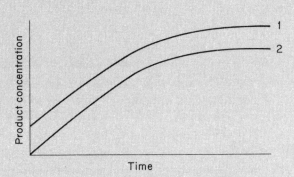

Fig. 6.1c.   *Reaction profiles for* (1) *a turbid sample, and* (2) *a normal, clear sample*

(*i*) Use this figure to describe one of the advantages of assay systems which depend upon the measurement of reaction rates during the early part of the reaction.
(*ii*) List the other advantages.

Response

(*i*) The numerical value of the equilibrium level is raised in the turbid sample and so the inclusion of appropriate blanks is essential to allow for this. Notice however that the slope of the early part of the graph is identical for both reactions, and hence the same reaction rate is obtained. The time of reaching the equilibrium plateau is the same in both cases and this parameter could be used to measure reaction rate if necessary.

(*ii*) — Decrease of time required for the assay.

— Maximum discrimination between different samples, interfering reactions etc.

— Maximum reproducibility and reliability of the assay.

****

SAQ 6.2a	Take each of the following brief descriptions of enzyme assay methods and identify the type of measurement being undertaken (one-point, two-point, fixed-time, variable-time etc). (*i*) *Alkaline phosphatase* Sample is incubated with phenolphthalein mono-phosphate. After 20 minutes 0.1 mol dm^3 pH 11.2 buffer solution is added to inactivate the enzyme and increase the colour intensity. Measurement is made at 550 nm against reagents processed without sample. (*ii*) *Caeruloplasmin I* Sample is incubated with o-dianisidine and portions are removed at 5 and 15 minute intervals and added to sulphuric acid. Absorbance measurements are made at 540 nm against water. $\longrightarrow$

SAQ 6.2a
(cont.)

(iii) Caeruloplasmin II
Sample is incubated with a substrate in a spectrophotometer cuvette and changes in absorbance at 460 nm are recorded on a chart. The reaction rate is calculated from the slope at zero time.

(iv) Lactate dehydrogenase
Sample is incubated with lactate and NAD^+ for 7 minutes. An acidic solution of cuprous (copper(I)), ions and neocuproine is added to stop the reaction and develop the colour. The absorbance is measured at 455 nm against a sample of serum.

(v) Aspartate aminotransferase
Sample is incubated with a complex substrate mixture in a spectrophotometer cuvette. A short time is allowed for the lag period to pass and the reaction rate analyser collects 5 absorbance measurements at 0.1 s intervals. This is repeated after 6 s and the difference between the two sets used for the rate calculation.

(vi) Alkaline phosphatase
Sample is incubated with disodium-p-nitrophenyl-phosphate and the enzymic release of p-nitrophenol is measured. The analyser is set to record the time when transmittance passes 90% and 88.2% values, and the time difference is used to calculate the reaction rate.

Response

(*i*)   A one-point assay with reagent blank.
(*ii*) A two-point, interrupted, fixed-time assay.
(*iii*) A simple continuous (kinetic) assay for the initial reaction rate,
(*iv*) A one-point assay with a sample blank.
(*v*)   A two-point continuous (kinetic) assay.
(*vi*) A two-point variable-time, fixed-signal assay.

****

SAQ 6.2b   List three potential problems with choosing time
                points t_1 and t_2 for the initial and final assay
                points in a typical two-point enzyme assay pro-
                cedure (Fig. 6.2b)

Response

Other samples might show lag periods, early high activity levels, or
may contain a high activity of the enzyme of interest. Points t_1 and
t_2 are too close to the ends of the reaction profile.

****

SAQ 6.2c   Graph (i) of Fig. 6.2c represents an enzyme reac-
                tion profile with a lag period and an equilibrium
                plateau. Graph (ii) represents a reaction without
                a lag period but which proceeds at the same rate.

                Use each graph to determine the reaction rate
                (1) after 4 minutes and (2) between 2 and 4 min-
                utes. Comment on the results obtained. What are
                the names given to the measurement approaches
                in (1) and (2)? $\longrightarrow$

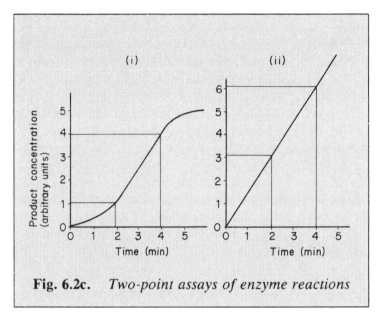

Fig. 6.2c. *Two-point assays of enzyme reactions*

Response

The values obtained are shown in the table:

	Rate min^{-1}	
	(i)	(ii)
(1)	1.0	1.5
(2)	1.5	1.5

Notice that the one-point method used in (1) gives different values for the two graphs, whereas the two-point method used in (2) gives identical ones.

****

SAQ 6.2d A highly active sample shows an apparent activity of $100 \ U \ cm^{-3}$, on dilution to 50% the activity becomes $40 \ U \ cm^{-3}$. Explain why the latter value might be obtained instead of $50 \ U \ cm^{-3}$.

Response

On dilution of a biological specimen, the enzyme becomes diluted, but so do other enzymes, cofactors, alternative substrates, inhibitors etc. It may be the case that important enzyme activators or required cofactors have been diluted so that a disproportionate reduction in rate is seen.

****

SAQ 6.2e List the advantages of centrifugal analysers for the assay of enzyme reaction rates.

Response

(*i*) Double beam spectrophotometer performance with only single beam construction reduces cost and some potential sources of error.

(*ii*) All reactions run in an exactly parallel time frame rather than sequentially. Thus exact timing of initiation is relatively unimportant and blanking is more accurate.

(*iii*) The collection of a considerable number of measurements at closely spaced time intervals beginning very early in the reaction, greatly improves the quality and reliability of the results and increases the speed of the assay.

****

> **SAQ 6.3a** Define or briefly describe the relevance of each of the following to reaction rate measurements:
>
> (*i*)   signal averaging and elimination;
> (*ii*) measurement of slopes;
> (*iii*) comparison of integral areas;
> (*iv*) scatter limits;
> (*v*)   derivative calculations.

Response

Details of all the above topics are included in the text but the following notes are given as a reminder.

(*i*)   Some analysis systems, particularly centrifugal analysers, collect so many data that memory overload is a possibility, and in any case there is little benefit in using them all. They usually therefore reduce the number by either ignoring some data or averaging them all.

(*ii*) The slope is directly related to the enzyme activity and can be used as the basis of the rate calculation. In addition by a comparison of the slopes obtained over short, adjacent regions, the linear portion of a reaction profile can be identified and a more reliable value for the rate obtained.

(*iii*) Instead of comparing slopes at selected points on the curve, a comparison can be made of areas under the curve (integrals). The difference between adjacent areas is proportional to the reaction rate.

(*iv*) When using statistical methods to fit the best line to the experimental data, the quality of the line can be adversely affected by points which are markedly adrift of it. Automatic analysers usually assign maximum allowable limits for the scatter of points and eliminate from the calculations those which lie outside these limits.

(*v*)   Derivative calculations are mathematical operations that transform the original data in such a way as to produce a marked change in their plot when the rate of change of the data itself alters. The beginning and end of the linear portions of a reaction rate graph are such points.

****

SAQ 7.2a	Distinguish between the following terms: (*i*) definitive and reference methods; (*ii*) optimised and recommended methods.

Response

(*i*)   Definitive method has no known sources of error, a reference method has the minimum error of those methods currently available.

(*ii*) An optimised method uses reaction conditions which have been shown to produce the maximum activity, a recommended method is one that is suitable for general laboratory use (in the sense that it has workable substrate concentrations, is reasonably cost effective etc). It may use different conditions from those in the optimised method.

****

SAQ 7.3a

> (*i*) Distinguish between the terms primary and secondary standards, and assigned and consensus values.
>
> (*ii*) State one advantage and one disadvantage of using animal sera instead of human sera for quality control of enzyme assays.

Response

(*i*) A primary standard has a precisely known composition (eg by virtue of being a weighed solution); a secondary standard has a composition which has been experimentally determined by comparison with other methods and standards.

An assigned value is one given to a standard usually by some professional body based upon their investigations or knowledge of it; a consensus value is given to a standard following a large number of analyses usually in a quality control programme.

(*ii*) Among the relatively few advantages are cost and freedom from human diseases such as infectious hepatitis.

The disadvantages derive from the fact that animal sera may not give quantitatively or qualitatively the same response as human material due to differences in:

— matrix effects arising from the interaction of components;

— actual enzyme composition, especially of isoenzymes;

— enzyme properties (kinetics, stability, specificity etc).

****

SAQ 7.3b   Explain why:

(*i*)   reconstituted sera are turbid in appearance;

(*ii*) reconstituted sera are different in their isoenzyme (ISE) pattern compared with normal sera;

(*iii*) supplemented sera are different in their isoenzyme pattern from normal sera.

Response

(*i*)   Some proteins, especially lipoproteins, denature on freezing and hence do not redissolve on reconstitution. The suspended particles give a turbid appearance.

(*ii*) For some enzymes the isoenzymic forms vary in their response to freezing. If some forms denature on freezing then the ISE pattern will be different on reconstitution or thawing. In addition, there is a tendency for the sub-units of the ISE to separate on freezing and randomly recombine later; this also changes the isoenzyme pattern.

(*iii*) Sera are frequently supplemented with extracts from specific organs which may have a particular ISE composition, thus affecting the pattern in the standard serum.

****

SAQ 7.3c

> List at least three problems with lyophilised sera for use as standards in quality control programmes.

Response

Among the problems you could have mentioned are

(*a*) the various changes occurring during processing, eg protein denaturation effects, changes in protein/protein and protein/other molecule interactions, and the variation among enzymes and especially isoenzymes in their response to these;

(*b*) production problems such as variations in sample vial filling and;

(*c*) user related problems such as errors due to variation in reconstitution, increase in cost due to wastage and variations in stability and activity of the enzymes following reconstitution are also relevant.

****

SAQ 8.4a

> (*i*) A large group of enzymes present in blood comprise those that are released from cells in various 'accidental' ways. Other enzymes in blood have definite functions there – state two distinct such functions.
>
> (*ii*) State two changes that can be experimentally demonstrated to occur in cell membranes as a result of diseases.
>
> (*iii*) Would an effect of a disease on DNA structure (and hence on the structure and production rate of enzymes) produce earlier or later changes in circulating enzyme level than an effect on the outer cell membrane? ⟶

SAQ 8.4a
(cont.)

(*iv*) Select from the following list the enzyme which is likely to be most easily released from cells as a result of disease.

(*a*) cell surface lipase;
(*b*) mitochondrial ATPase;
(*c*) cytoplasmic aldolase;
(*d*) Golgi body glycosyl transferase;
(*e*) nuclear RNA polymerase.

(*v*) Can you give two examples of different ways in which the assay technique for an enzyme can affect its reference range.

(*vi*) Choose one of the enzymes from the following list which has a normal blood activity showing a relationship to body weight of the individual.

LDH, AP, GGT, AM, CK, GOT, G-6-PD.

(*vii*) List with examples three physiological states that can alter a patient's enzyme activity from the reference range of the whole population in the absence of disease.

(*viii*) On average which of the following ratios could one expect to find for the cellular: blood activities of a typical enzyme?

$10^6 : 1, 10^5 : 1, 10^3 : 1, 10^2 : 1, 10^{-2} : 1$

(*ix*) Diseases can result in a change in the basal level of an enzyme by increasing the leakage from damaged or dying cells. State two other distinct ways in which diseases can change basal levels.

Response

(*i*)   Perhaps the most well known group of enzymes with functions in the blood are those involved in defence (clotting and anti-microbial activity), and in metabolism (frequently of food materials such as lipids).

(*ii*)   You could have selected your answers from visible changes in width, staining properties etc, or changes in measured physical factors such as electrical charge, conductance etc.

(*iii*)   An effect on the outer membrane could result in a release of existing enzyme immediately; it would be some hours before an effect on DNA produced a significant change, since the cell already has a full complement of enzymes in the cytoplasm.

(*iv*)   Providing no other factor is involved (which of course is unlikely) then the nearer an enzyme is to the cell surface and the less it is integrated with other structures (multi-enzyme complexes, membranes etc) the more likely it is to be released.

Thus (*a*) the surface lipase ought in principle to be easily released and indeed in practice it is released by many agents including the anti-clotting factor, heparin. The cytoplasmic aldolase (*c*) as a relatively soluble free enzyme ought also to be released comparatively easily compared with the organelle bound ones (*b*), (*d*), (*e*), and compared also with other cytoplasmic enzymes such as glycogen synthetase which happens to be associated with large particles in the cytoplasm.

(*v*)   Sample problems are a major source of variation and range from those associated with the site (venous or capillary) and method of sampling (prick and droplet or large volume; tourniquet or free draining etc) to storage problems (particularly temperature and the use of chemical preservatives).

Variations resulting from different assay methods are legion and include direction of reaction chosen, reaction rate or end product approach, substrate selected, reaction conditions used and the choice of instrument for analysis of the reaction.

(*vi*) Muscle based enzymes are perhaps the best examples and LDH, CK, and AST are particularly noteworthy.

(*vii*) You could have chosen from meals, malnutrition, drug therapy, rhythms, pregnancy, muscular activity, posture.

(*viii*) $10^2 : 1$ to $10^4 : 1$ are the ratios you find for the majority of enzymes so you could have chosen $10^3 : 1$ and $10^2 : 1$ from the list.

(*ix*) Cases in which a decrease in leakage occurs are rare if known at all, but increases and especially decreases in rate of synthesis are common, as are increases and particularly decreases in cell number.

****

SAQ 8.4b	The myocardial infarction (MI) is a condition that develops very quickly, whereas many types of liver hepatitis are due to infections and develop more slowly. The enzyme profile in serum following an MI is very similar to that of the heart muscle whereas the correlation is much poorer for hepatitis. What might be the cause of this difference?

Response

Many of the factors discussed in the text, eg differential rates of release of the various liver enzymes, differential diffusion in the intracellular fluid and transfer to the capillaries, could be of significance but the most likely factor is the different rates of degradation of serum enzymes. With the serum enzymes being released over a long period of time many more molecules of the short half-life forms will have been degraded compared with the more stable long half-life types. A distortion of the pattern is therefore developed.

****

SAQ 8.4c

An enzyme is released from a damaged organ at a rate of 100 units per day but it is degraded at a rate of 10% of its activity per day. For simplicity we will assume the processes occur stepwise (which of course is not the case at all), and the following changes will then be found.

Time (days)	Expected activity (100 U day^{-1} released)	Actual observed activity (10% degraded per day)	Actual observed activity (30% degraded per day)
1	100	100–10 = 90	
2	200	190–19 = 171	
3	300	271–27 = 244	
4	400	344–34 = 310	
5	500	410–41 = 369	
6	600	469–47 = 422	
7	700	522–52 = 470	

Thus the level after one week will be 470 units rather than 700. Carry out the same calculation assuming a 30% degradation rate and comment on the result.

Response

The calculation is as follows

Time (days)	Expected activity (100 U day^{-1} released)	Actual observed activity (10% degraded per day)	Actual observed activity (30% degraded per day)
1	100	100–10 = 90	100–30 = 70
2	200	190–19 = 171	170–51 = 119
3	300	271–27 = 244	219–66 = 153
4	400	344–34 = 310	253–76 = 177
5	500	410–41 = 369	277–83 = 194
6	600	469–47 = 422	294–88 = 216
7	700	522–52 = 470	316–95 = 221

With a degradation rate of 30% the level after 1 week is only 221 units which is less than half that obtained with a 10% degradation rate. The importance of the degradation process in influencing the enzyme level available for assay is well illustrated by this calculation.

****

SAQ 8.5a

In liver cells the AST activity is approximately 60% higher than the ALT activity. However after a period of acute hepatitis the serum activity of AST is much less than that of ALT.

One of the causes of such a difference could well be a differential rate of inactivation of the two enzymes; but bearing in mind the cellular location of the two enzymes, what other explanation is possible?

Response

40% of the liver cell AST is mitochondrial and would be slow to be released. In acute hepatitis the serum pattern resembles reasonably closely the cellular pattern if this 40% AST is deducted.

****

SAQ 8.5b

Some assays on a particular patient gave the following values for the serum enzyme activities.

Enzyme abbreviation	Enzyme activity $U\ dm^{-3}$	Reference range $U\ dm^{-3}$
AST	50	5–20
ALT	20	5–25
CK	15	0–50
LDH	150	80–240

(*i*) An acid phosphatase level of 500 U dm^{-3} (with a reference range of 10–170), would suggest the possibility of a liver disease. What data from the above table would tell you that the origin of the disease is not in fact the liver?

(*ii*) If the liver were the site of the damage would you expect the AST/GGT ratio to be high or moderate?

(*iii*) What other approach could be made to exclude the liver as the site of disease?

Response

(*i*)   A significant disease of the liver would be expected to release a reasonable amount of ALT. The origin of this patient's disease lies elsewhere therefore, probably in the bones.

(*ii*) Moderate, since the GGT is relatively specific to the liver and would probably be released along with the AST.

(*iii*) A study of the ISEs of LDH would be useful since certain forms are specific to the liver and would be released if this organ were diseased (see also 9.1).

****

SAQ 9.2a

(*i*)   List the major roles of the liver.

(*ii*)   What is a xenobiotic?

(*iii*) What is the major chemical product of the degradation of haemoglobin?

(*iv*)   Distinguish the three major types of jaundice.

(*v*)   5'-Nucleotidase is very specific to the liver; why is it not used more commonly in diagnosis?

(*vi*)   State two distinct conditions that can result in an increase in LDH 4 + 5 activity.

(*vii*) Which enzyme shows a substantial rise as a result of persistent alcoholism? State one reason why correlating an increase in this enzyme with alcoholism needs to be done with care.

Response

(*i*) — Interconversion of the products of digestion;
— storage of materials;
— plasma protein synthesis;
— metabolism of xenobiotics;
— red blood cell destruction;
— haemoglobin degradation.

(*ii*) A compound foreign to the body.

(*iii*) Bilirubin.

(*iv*) Haemolytic (pre-hepatic) due to a high rate of red blood cell breakdown, hepatocellular (hepatic) due to liver or bile system metabolic defects, obstructive (post-hepatic) due to blockage in the bile system.

(*v*) The assay is technically difficult and relatively insensitive.

(*vi*) — Effects on liver parenchymal cells;
— cancers;
— bone joint inflammation;
— white blood cell destruction.

(*vii*) GGT. Synthesis of the enzyme can be induced by drugs resulting in increase in circulating level.

****

SAQ 9.2b

Two individuals with (*i*) haemolysis and (*ii*) an inherited metabolic defect of the liver called Gilbert's Disease, had the following serum enzyme activities. Normal values are given in brackets. Explain these results.

		(*i*)	(*ii*)
Bilirubin	(<20 μmol dm^{-3})	54	42
Albumin	(30–50 g dm^{-3})	40	46
ALP	(30–120 U dm^{-3})	106	30
ALT	(<35 U dm^{-3})	14	30

Response

In neither of the above conditions are the liver cells primarily affected and hence enzyme release does not occur. The cells may be affected later in the course of the disease, with the exact timing of this depending on its precise nature and severity.

****

SAQ 9.2c

In cases of pre-hepatic (haemolytic) jaundice why is an increase in LDH commonly found? How could you confirm the probable origin of this enzyme?

Response

The enzyme could possibly have come from the liver but, at least in the early stages and in mild conditions, the liver cells tend not to be affected since the condition is primarily due to defective red blood cells. The LDH would probably have come from bursting red blood cells and could be confirmed as such by identification as LDH type 1. The liver would produce types 4 + 5.

****

SAQ 9.2d	The figures in column (*i*) of the following table were found in a typical case of obstructive jaundice analysed soon after the condition developed. When the condition was left untreated the values in column (*ii*) were obtained. Explain the differences shown.	

		(*i*)	(*ii*)
Bilirubin	($< 20\ \mu\text{mol dm}^{-3}$)	150	200
Albumin	($30\text{--}50\ \text{g dm}^{-3}$)	48	30
ALP	($30\text{--}120\ \text{U dm}^{-3}$)	570	565
ALT	($< 35\ \text{U dm}^{-3}$)	30	363

Response

It is the case that very often conditions in one organ system will adversely affect other organs either immediately or in the longer term. In this case the persistent back-pressure due to bile accumulation and the toxic affects of the bile itself cause hepatocyte damage resulting in an increase in ALT released from the liver parenchyma cells.

****

SAQ 9.2e | The table shows some typical enzyme activities for a case of acute viral hepatitis with column (*i*) obtained 1 day after the patient began to feel unwell, and column (*ii*) obtained 4 weeks after infection, by which time he had apparently recovered.

		(*i*)	(*ii*)
Bilirubin	($<$20 mmol dm^{-3})	30	18
Albumin	(30–50 g dm^{-3})	43	38
AP	(30–120 U dm^{-3})	118	110
ALT	($<$35 U dm^{-3})	1140	32

What is the significance, with regard to ease of diagnosis, of the low bilirubin and high ALT activities in column (*i*)?

What is the significance of the figures in column (*ii*)?

Response

The low bilirubin level will mean that jaundice will not be visible and although the patient may feel unwell, diagnosis will not be particularly easy. The high ALT value produced so early in the development of the condition will be of great value in narrowing down the causes of the illness to the liver.

The serum values have returned to normal within a 3–4 week period, confirming the recovery made by the patient.

****

SAQ 9.3a

A measurement of total CK activity was undertaken 1.5 days after a probable MI.

You would like to confirm the time of onset of the MI – what other measurement could you carry out?

The patient is very ill and you would like to monitor the possibility of repeat infarcts over the next few days. How could you do it?

Response

It is necessary to determine the position of the value you have obtained on the enzyme profile by carrying out a measurement after a suitable time lapse. Whereas the total CK measurement at 2–3 days might be useful, the removal of the first sample and the general muscle trauma associated with an MI, might have raised the CK level due to the release of CK MM. A series of sequential measurements of the CK MB level might provide the information, or measurement of a rising enzyme such as LDH might be of value.

The most sensitive method for detecting secondary infarcts is by assaying the CK MB isoenzyme as discussed in the text.

****

SAQ 9.3b

There has been a time lapse of 10 h since the onset of chest pain. Out of the following list which pair of enzymes would be most appropriate to determine whether the pain was due to an MI?

(*i*) CK and LDH1; $\longrightarrow$

SAQ 9.3b
(cont.)

> (*ii*) CK and AST;
>
> (*iii*) CK and ALT;
>
> (*iv*) AST and ALT;
>
> (*v*) AST and LDH1.

Response

Since ALT is not released to any great extent by an MI, (*iii*) and (*iv*) are of little use except for the exclusion of other causes. LDH is released comparatively slowly, and peaks after CK and AST are nearly back to normal, hence (*i*) and (*v*) are not particularly useful. The most suitable combination would therefore be (*ii*).

****

SAQ 9.3c

> There has been a time lapse of about 3 days since the onset of chest pain. Which enzyme(s) from the list below would you expect to see raised if (*i*) the pain was due to as MI, and (*ii*) it was due to angina.
>
> ALT,
> AST,
> CK MB,
> Total CK,
> LDH1.

Response

(*i*) By 3 days the early rising enzymes such as AST and CK MB will have fallen back towards normal. Total CK activity will be falling but may still be above normal. ALT is not released to any significant extent following heart trauma. LDH1 activity should be rising and be easily demonstrated.

(*ii*) Little change in any enzyme is to be expected although AST may occasionally be slightly raised.

****

Units of Measurement

For historic reasons a number of different units of measurement have evolved to express quantity of the same thing. In the 1960s, many international scientific bodies recommended the standardisation of names and symbols and the adoption universally of a coherent set of units—the SI units (Système Internationale d'Unités)—based on the definition of five basic units: metre (m); kilogram (kg); second (s); ampere (A); mole (mol); and candela (cd).

The earlier literature references and some of the older text books, naturally use the older units. Even now many practicing scientists have not adopted the SI unit as their working unit. It is therefore necessary to know of the older units and be able to interconvert with SI units.

In this series of texts SI units are used as standard practice. However in areas of activity where their use has not become general practice, eg biologically based laboratories, the earlier defined units are used. This is explained in the study guide to each unit.

Table 1 shows some symbols and abbreviations commonly used in analytical chemistry; Table 5 relates the many systems of naming enzymes and Table 6 is a glossary of abbreviations used in this particular text. Table 2 shows some of the alternative methods for expressing the values of physical quantities and the relationship to the value in SI units.

More details and definition of other units may be found in the *Manual of Symbols and Terminology for Physicochemical Quantities and Units*, Whiffen, 1979, Pergamon Press.

Table 1   *Symbols and Abbreviations Commonly used in Analytical Chemistry*

Å	Angstrom
$A_r(X)$	relative atomic mass of X
A	ampere
E or U	energy
G	Gibbs free energy (function)
H	enthalpy
J	joule
K	kelvin (273.15 + $t\,°C$)
K	equilibrium constant (with subscripts p, c, therm etc.)
K_a, K_b	acid and base ionisation constants
$M_r(X)$	relative molecular mass of X
N	newton (SI unit of force)
P	total pressure
s	standard deviation
T	temperature / K
V	volume
V	volt ($J\ A^{-1}\ s^{-1}$)
$a, a(A)$	activity, activity of A
c	concentration / mol dm^{-3}
e	electron
g	gramme
i	current
s	second
t	temperature / °C
bp	boiling point
fp	freezing point
mp	melting point
$\approx$	approximately equal to
$<$	less than
$>$	greater than
e, $\exp(x)$	exponential of x
$\ln x$	natural logarithm of x; $\ln x = 2.303 \log x$
$\log x$	common logarithm of x to base 10

Table 2 *Alternative Methods of Expressing Various Physical Quantities*

1. **Mass (SI unit : kg)**

$$g = 10^{-3} \text{ kg}$$
$$mg = 10^{-3} \text{ g} = 10^{-6} \text{ kg}$$
$$\mu g = 10^{-6} \text{ g} = 10^{-9} \text{ kg}$$

2. **Length (SI unit : m)**

$$cm = 10^{-2} \text{ m}$$
$$\text{Å} = 10^{-10} \text{ m}$$
$$nm = 10^{-9} \text{ m} = 10\text{Å}$$
$$pm = 10^{-12} \text{ m} = 10^{-2} \text{ Å}$$

3. **Volume (SI unit : m^3)**

$$l = dm^3 = 10^{-3} \text{ m}^3$$
$$ml = cm^3 = 10^{-6} \text{ m}^3$$
$$\mu l = 10^{-3} \text{ cm}^3$$

4. **Concentration (SI units : $mol \ m^{-3}$)**

$$M = mol \ l^{-1} = mol \ dm^{-3} = 10^3 \ mol \ m^{-3}$$
$$mg \ l^{-1} = \mu g \ cm^{-3} = ppm = 10^{-3} \ g \ dm^{-3}$$
$$\mu g \ g^{-1} = ppm = 10^{-6} \ g \ g^{-1}$$
$$ng \ cm^{-3} = 10^{-6} \ g \ dm^{-3}$$
$$ng \ dm^{-3} = pg \ cm^{-3}$$
$$pg \ g^{-1} = ppb = 10^{-12} \ g \ g^{-1}$$
$$mg\% = 10^{-2} \ g \ dm^{-3}$$
$$\mu g\% = 10^{-5} \ g \ dm^{-3}$$

5. **Pressure (SI unit : $N \ m^{-2} = kg \ m^{-1} \ s^{-2}$)**

$$Pa = Nm^{-2}$$
$$atmos = 101 \ 325 \ N \ m^{-2}$$
$$bar = 10^5 \ N \ m^{-2}$$
$$torr = mmHg = 133.322 \ N \ m^{-2}$$

6. **Energy (SI unit : $J = kg \ m^2 \ s^{-2}$)**

$$cal = 4.184 \text{ J}$$
$$erg = 10^{-7} \text{ J}$$
$$eV = 1.602 \times 10^{-19} \text{ J}$$

Table 3 *Prefixes for SI Units*

Fraction	Prefix	Symbol
10^{-1}	deci	d
10^{-2}	centi	c
10^{-3}	milli	m
10^{-6}	micro	μ
10^{-9}	nano	n
10^{-12}	pico	p
10^{-15}	femto	f
10^{-18}	atto	a

Multiple	Prefix	Symbol
10	deka	da
10^2	hecto	h
10^3	kilo	k
10^6	mega	M
10^9	giga	G
10^{12}	tera	T
10^{15}	peta	P
10^{18}	exa	E

Table 4   *Recommended Values of Physical Constants*

Physical constant	Symbol	Value
acceleration due to gravity	g	9.81 m s^{-2}
Avogadro constant	N_A	$6.022\,05 \times 10^{23}$ mol^{-1}
Boltzmann constant	k	$1.380\,66 \times 10^{-23}$ J K^{-1}
charge to mass ratio	e/m	$1.758\,796 \times 10^{11}$ C kg^{-1}
electronic charge	e	$1.602\,19 \times 10^{-19}$ C
Faraday constant	F	$9.648\,46 \times 10^{4}$ C mol^{-1}
gas constant	R	8.314 J K^{-1} mol^{-1}
'ice-point' temperature	T_{ice}	273.150 K exactly
molar volume of ideal gas (stp)	V_m	$2.241\,38 \times 10^{-2}$ m^3 mol^{-1}
permittivity of a vacuum	ϵ_0	$8.854\,188 \times 10^{-12}$ kg^{-1} m^{-3} s^4 A^2 (F m^{-1})
Planck constant	h	$6.626\,2 \times 10^{-34}$ J s
standard atmosphere pressure	p	$101\,325$ N m^{-2} exactly
atomic mass unit	m_u	$1.660\,566 \times 10^{-27}$ kg
speed of light in a vacuum	c	$2.997\,925 \times 10^{8}$ m s^{-1}

Table 5 *Enzyme Nomenclature*

Trivial Name	Common Abbreviation	Systematic Name	Enzyme Commission Number
Acid phosphatase	AP	Orthophosphoric monoester phosphohydrolase	3.1.3.2
Adenylate kinase	AK	ATP: AMP phosphotransferase	2.7.4.3
Alanine transaminase	ALT (GPT)	L-alanine: 2-oxoglutarate aminotransferase	2.6.1.2
Alkaline phosphatase	ALP	Orthophosphoric monoester phosphohydrolase	3.1.3.1
Amylase	AM	1,4-α-D-glucanohydrolase	3.2.1.1
Aspartate transaminase	AST (GOT)	L-aspartate: 2-oxoglutarate aminotransferase	2.6.1.1
ATPase	ATPase	ATP phosphohydrolase	3.6.1.3
Cholinesterase	CHE	Acetylcholine hydrolase	3.1.1.7
Creatine kinase	CK	ATP: creatine phosphotransferase	2.7.3.2
Gamma glutamyl transferase	GGT	γ-glutamyl transferase	2.3.2.2
Glucose-6-phosphate dehydrogenase	GPD	D-glucose-6-phosphate: NADP$^+$-1-oxidoreductase	1.1.1.49
Glutamate dehydrogenase	GLDH	L-glutamate: NAD$^+$ oxidoreductase	1.4.1.2
Hydroxybutyrate dehydrogenase	HBDH	As lactate dehydrogenase	1.1.1.27
Isocitrate dehydrogenase	ICDH	Threo-Ds-Isocitrate: NADP$^+$ oxidoreductase	1.1.1.42
Lactate dehydrogenase	LDH	L-lactate: NAD$^+$ oxidoreductase	1.1.1.27
Lecithin: cholesterol acyltransferase	LCAT	Lecithin: cholesterol acyltransferase	2.3.1.43
Lipase	LP	Glycerol ester hydrolase	3.1.1.3
5'-nucleotidase	5NT	5'-Ribonucleotide phosphohydrolase	3.1.3.5
Pyruvate kinase	PK	ATP: pyruvate phosphotransferase	2.7.1.40
Sorbitol dehydrogenase	SDH	L-iditol: NAD$^+$ oxidoreductase	1.1.1.14

Table 6   *Other Abbreviations and Acronyms*
used in This Unit

A	*absorbance*
A_{340}	absorbance at a fixed wavelength (340 nm)
ADP	adenosine diphosphate
AMP	adenosine monophosphate
ATP	adenosine triphosphate
CSF	cerebrospinal fluid
DNPH	dinitrophenylhydrazine
EDTA	ethylenediamine tetraacetic (ethanoic) acid
ECG	electrocardiograph
FMN	flavin mononucleotide
ISE	isoenzyme
NAD^+	nicotinamide adenine dinucleotide
NADH	reduced form of NAD^+
$NADP^+$	nicotinamide adenine dinucleotide phosphate
PMS	phenazine methosulphate
RMM, M_r	relative molecular mass
T	temperature, K
t	time
TRIS	tris(hydroxymethyl)methylamine
U	International unit of enzyme activity

Specific References

Brown A., Mitchell F. L. and Young D. S., *Chemical Diagnosis of Disease*, Elsevier, 1982.

Burtis C. A. and Mrochek J., Data Processing for Centrifugal Analysis in *Centrifugal Analysers in Clinical Chemistry*, ed. Price C. P. and Spencer K., Praeger, 1980.

Elin R. and Gray B., *Clinical Chemistry* (1984), **30**, 129–131.

Holme D. and Peck H., *Analytical Biochemistry*, Longman, 1983.

Holton J. B., Prenatal Diagnosis in Inherited Disease in *Laboratory Investigations of Fetal Disease*, ed Barson A. and Davis J., Wright, 1981.

IFCC Expert Panel on Enzymes, *Provisional Recommendations on IFCC Methods for the Measurement of Catalytic Concentrations of Enzymes*, Part 2—IFCC Method for Aspartate Aminotransferase. *Clin. Chem. Acta* (1976), **70**, F19–42.

Jung K., Schrieber G. and Grutzmann K., *Clin. Chem. Acta* (1982), **120**, 367–371.

Krause R., *Clinical Chemistry* (1974), **20**, 775–782.

Latt S. A. and Darlington G. J., ed. Prenatal Diagnosis: Cell Biological Approaches in *Methods in Cell Biology*, Vol. 26. Academic Press, 1982.

Lodensen J. H., Non-analytical Sources of Variation in Clinical Chemistry Results in *Gradwohl's Clinical Laboratory Methods and Diagnosis*, ed Sonnewirth A. and Jarett L., Mosby, 1980.

London J., *Clinical Chemistry* (1975), **21**, 1939–1952.

Martin H., Gudzinowicz B. and Fanger C., *Normal Values in Clinical Chemistry*, Dekker, 1978.

Roberts R., Ahuymada G. and Sobel B., *Estimation of Infarct Size*, Upjohn Co., 1975.

Rosalki S. (1980), *Ann. Clin. Biochem.*, **17**, 74–77.

Schmidt E. and Schmidt F., *Diagnostik* (1975), **8**, 427–432.

Schmidt E. and Schmidt F., *Brief Guide to Practical Enzyme Diagnosis* 2nd ed, Boehringer Mannheim GmbH, 1976.

Siggaard-Andersen S., Electrochemistry in *Textbook of Clinical Chemitry*, ed Tietz N., Saunders, 1986.

Skoog D. and West D., *Principles of Instrumental Analysis*, 2nd ed, Holt Saunders, 1985.

Steinhausen R. and Price C., Principles and Practices of Dry Chemistry Systems in *Recent Advances in Clinical Biochemistry*, **3**, ed Price C. P. and Alberti K. G. M. M., Churchill Livingston, 1985.

Tanishma K., *Clinical Chemistry* (1977), **23**, 1873–1877.

Tietz N., *Clinical Chemistry* (1983), **29**, 751–761.

Vadgama P. and Davis G., Biosensors in Clinical Chemistry in *Medical Laboratory Sciences* (1985), **42**, 333–345.

Varley H., Gowenlock A. and Bell M., *Practical Clinical Biochemistry*, Vol I and II, 5th ed, Heinemann, 1980.

Wilkinson T., *The Principles and Practice of Diagnostic Enzymology*, Arnold, 1978.

Williams B. and Goulding K., *Biologist's Guide to the Principles and Techniques of Practical Biochemistry*, 3rd ed, Arnold, 1986.

Wooton I. and Freeman H., *Microanalysis in Medical Chemistry*, 6th ed, Churchill, 1982.

Wroblewski F., *Adv. in Clin. Chem.* (1958), **1**, 340–352.

Young E., Willcox P., Whitfield A. and Patrick A. J., *Med. Gen.* (1975), **12**, 224–229.